FIRST LIGHT

365 Daily Devotionals

EVANGELIST WIL RICE IV

First Light

© 2016 Bill Rice Ranch, Inc. All rights reserved.

Scripture quotations are taken from the *Holy Bible*, King James Version

Printed in the United States of America

No part of this publication may be reproduced, stored in a retrieval system, or transmitted in any form or by any means—electronic, mechanical, photocopy, recording, or any other—without the prior written permission of the publisher. The only exception is brief quotations in printed reviews.

ACKNOWLEDGEMENTS

First Light is a compilation of some of the weekly devotionals the Ranch has sent to email subscribers. The devotionals in this book have been recorded from the morning devotions at the Bill Rice Ranch and then reproduced in print by the diligent work of Susanna Flanders, Jake Griffith, and Rebekah McConnell. They have spent countless hours listening to these daily Bible challenges and many more hours transcribing, editing, and distributing them. There are those who could have said it better than I have, but I could not have asked for better help than they have given. I hope that this book will encourage you to read your Bible and to think as you read!

With a warm handshake,

Wil Rice IV

IN A PERFECT WORLD...

DAY 1

Genesis 1:1 In the beginning God created the heaven and the earth.

What would a perfect world be like? In Scripture, we get glimpses of a perfect world at the beginning in Genesis and at the end in Revelation. In Genesis, we see what the world *was* like at Creation, before the fall of man and before the curse. In Revelation, we see what the world *will be* like someday when God creates a "new heaven and a new earth."

Are there any similarities between the two? Well, among others, the glaring similarity is that **both perfect worlds are perfect because they remain just the way God made them!** There is no sin, and no individuals are demanding their own way over God's. And guess what? In both perfect worlds, everything runs smoothly in perfect harmony. We can summarize the whole first chapter of Genesis with eight words: *"God created…it was so…it was good."*

Everything that God does is "good" just the way he does it. **Nothing that comes from God's hand ever needs improvement or adjustment.** That may seem like an obvious truth to emphasize, yet many of us live each day as if God's plan for us needs some manipulation and improvement. We want to order our days according to what *we* think is important and according to what our *own* ideals are.

If you take a few minutes to watch the news on any given day, you will see that mankind is completely inept! We vote for candidates that we think will solve all of our problems, and then we find that nothing really changes when they take office. And every day provides plenty of other examples of mankind messing things up! Take a look at the first perfect world. What messed things up? Was it God's plan that eventually went awry? No! **Things went terribly wrong when man started making decisions independently of God.**

So what does this mean for you today? You should set out to **order your day** according to God's ideals. You should **submit**

to God's Word today, even in the small areas. When you truly believe that God's way is best, you will submit, even when you don't understand. Don't manipulate your life to fit your own ideals, because you won't succeed at making it turn out right.

When the day is all done, who will be pleased with your decisions? Start your day praying, "Lord, you are the Creator; I am the creation. Help me to depend upon you for all that I need, and help me to submit to your plan for me." Trust God and obey—this is the way to get a glimpse of what that perfect world will be like someday.

DAY 2

GOD REMEMBERS YOU

Genesis 8:1 And God remembered Noah, and every living thing, and all the cattle that was with him in the ark: and God made a wind to pass over the earth, and the waters assuaged.

Memory is a powerful thing even when a person doesn't have a powerful memory. For instance, a man may not recollect events or the names of people, but if he remembers his wedding anniversary, that is powerful. It is one thing for people to remember us when they have very little to remember and there isn't much competition. It is another thing to realize that a God who rules the universe and has been in existence from eternity past remembers not just a century or an era, but **He remembers you** and me.

The Bible says in Genesis 8:1, *"And God remembered Noah."* God had judged the entire world with a universal flood after He had given people many years to make things right with Him. Noah had been a preacher of righteousness for generations, and yet the people of the world in Noah's day had turned their backs on God and seemed to shake their fists in His face.

Noah was utterly alone. He was the only man who had a family that loved God in the entire world. I remember my grandfather used to say, "I wish just once I could vote with the majority on something." Maybe you feel that way— like you are in a small minority. Maybe you feel as if no one in your world cares about God or about doing what is right.

Imagine Noah who literally lived in a world occupied by no one, other than his own family, who loved or knew God. Then imagine the fear of being overwhelmed when God judged with ferocity the whole earth. Noah was in a boat on an infinite ocean, alone in doing what is right. Noah was alone in his obedience and overwhelmed by God's righteousness, and God remembered Noah.

God remembers you even when you feel alone in your obedience and overwhelmed by God's righteousness. God, Who remembers everything that has ever happened and everyone who has ever lived, also remembers you because He cares.

BE THE HERO OF YOUR HOUSE, NOT OF THE WORLD

DAY 3

Genesis 9:21 And he drank of the wine, and was drunken; and he was uncovered within his tent.

I don't care who you are, where you live, or how much money you make, there will always be someone who will be unimpressed by what you are and what you do. That drives some people absolutely crazy. It leads many others to live lives that are not true, to do things that are not honest, or to portray themselves to be someone that they are not.

When I think about Noah, I think about a great man. He could almost be considered the hero of the world. All of us are descendants of one man who had the courage and determination to obey God and do what is right. Noah led his family. But like you and me, Noah was imperfect. Noah was largely defined, not by the great successes, but by the little defeats and victories of daily life.

After the floodwaters had receded and there was a new world populated only by Noah and his family, the Bible says in Genesis

9:20-21, *"Noah began to be an husbandman, and planted a vineyard. And he drank of the wine, and was drunken; and was uncovered within his tent."* What follows is a man who shamed himself in front of his children and was defeated in the small moments of life. It is a reminder that even the wisest of men can act foolishly when they are acting under the influence of something other than God.

It is better to be honored by the people in your house than to be the hero of the world. I hope you accomplish great things today and that God uses you in a wonderful way; but many times the moments that define our lives the most are the small, quiet moments shared by just us and those closest to us. It is those moments that oftentimes will define who we are and the legacy we leave.

Noah was a great and godly man; but like us, Noah was just human. He was flawed. He reminds us that whoever we are and whatever we do, nothing is more important than the integrity we show in the small moments of our lives.

◇◇◇

DAY 4
WHAT'S YOUR PRICE?

Genesis 12:10 And there was a famine in the land: and Abram went down into Egypt to sojourn there; for the famine was grievous in the land.

If you could either owe someone who didn't know or care about you or someone who knew you, loved you, and cared about you, which would you prefer? The answer is obvious, isn't it? It may be obvious, but this decision is not always easy in life.

Abram needed Egypt, went to Egypt, borrowed from Egypt, and owed Egypt. It's never a good idea to owe someone of low character like Pharaoh! Abram put himself in a position in which he felt the need to lie to preserve his life, at the risk of losing his marriage. He must have been one hungry man to put himself in that position!

We, too, often find a great need in our lives and look to the wrong people for the answer. When we do this, we are indebted to the wrong person, just like Abram. How much do you care about having the favor of people who don't care about having the favor of God? What is *your* price? What is your life worth?

Your actions will match the character of the person you are trying to please. Who do you owe? Who are your trying to please? What is most important to you? God could've taken care of Abram's need without jeopardizing his marriage and family. You can trust God with your life, your family, and your marriage too. What is your integrity worth?

◇◇◇

CHOICES

DAY 5

Genesis 13:11 Then Lot chose him all the plain of Jordan; and Lot journeyed east: and they separated themselves the one from the other.

If you had to choose between an easy choice and a hard choice, which one would you choose? That seems like a silly question, doesn't it? No one would prefer a hard choice.

What is the hardest choice you have ever made? I do not mean hard as in "not clear," but hard in the sense of a clear choice with unpleasant results. Abram and Lot were pushed to a tough decision because of possessions, but the results of the choice were suffered by people. Possessions, places, and people were all affected by this choice!

What seemed like an easy choice for Lot between the "good land" and the "bad land" actually turned out to be a hard choice for his family. **Sometimes the easiest choices to make are the hardest to live with.** Easy choices for a twenty-year-old "sowing his wild oats" become hard to live with when he's fifty. Hard choices for a twenty-year-old who wants to do right become easy when he's fifty.

Ask yourself these questions: "To what am I drawn?" and "To what am I heading?" I have a tendency to like quick and easy. How about you? We will all be faced with easy and hard choices today. Centuries have paid the price for what seemed like a very easy choice for Lot. What is your priority?

◇◇◇

DAY 6

IT STARTS AT HOME

Genesis 19:14 And Lot went out, and spake unto his sons in law, which married his daughters, and said, Up, get you out of this place; for the LORD will destroy this city. But he seemed as one that mocked unto his sons in law.

Have you ever known someone who was impossible to take seriously? Maybe he was such a jokester that every time he said something, you just assumed he was not serious. Other times we don't take a person seriously because he says something that is completely counter to the character that we know him to possess. It's not consistent. That is exactly what happened to Lot when God was about to judge Sodom and had provided mercy for him to escape with his family.

Genesis 19:14 says, *"And Lot went out, and spake unto his sons in law, which married his daughters, and said, Up, get you out of this place; for the LORD will destroy this city. But he seemed as one that mocked unto his sons in law."* Now sometimes we get the idea that his sons-in-law mocked Lot, but that is not what happened. Lot's sons-in-law thought Lot was mocking them. Lot had lived in such a way that when he came to them and said, "Get out of the city before God judges it," they thought he was joking.

The Bible tells us in the New Testament that Lot was righteous. Before God by position, Lot was a just and righteous man. But nobody would have guessed Lot to be righteous because no one knows the heart of another. All we know is what we see, and all that could be seen of Lot was evil.

God spared Lot because of what He saw based on the merits of the Messiah that would come, but God judged Lot's family. His family perished because of what they saw in Lot, a man who lived just like everyone else. Lot escaped the brimstone but not the destruction.

The point is that **a believer who does not live righteously in the world will not be taken seriously at home.** I want to be thought of well by people outside the confines of my house; but if I am going to do right, I need to live right every day in front of the people who see me and know me the best.

◇◇◇

JUST TELL THE TRUTH

DAY 7

Genesis 20:12 And yet indeed she is my sister; she is the daughter of my father, but not the daughter of my mother; and she became my wife.

Isn't it amazing that the same problems seem to snare us over and again? Sometimes, it's not as if the devil needs an entire arsenal of traps to trip us up. He just sends the same problem over and over again. Even a great man like Abraham, a man of faith, fell to the same sin on more than one occasion. He lied about his wife and said she was his sister for fear that she would be taken by some king, and he would be killed.

In Genesis 20, Abimelech had been told by Abraham that Sarah was his sister. So, he took Sarah into his harem. God had sent Abimelech a vision and told Abimelech he was a dead man because he had another man's wife. Abimelech responds in verse 5, *"In the integrity of my heart and innocency of my hands have I done this."* God says in verse 6, *"Yea, I know that thou didst this in the integrity of thy heart; for I also withheld thee from sinning against me: therefore suffered I thee not to touch her."* Because this man's heart was mainly right in this matter, God actually protected him from sinning and making it worse.

In contrast to this pagan king who had integrity of heart and innocence of hands, you have Abraham. Abraham protested when confronted with his lie and said in verse 12, *"Indeed she is my sister; she is the daughter of my father, but not the daughter of my mother; and she became my wife."* Now, that is not simple. That is not clear. That is not whole.

There is a difference between Abraham's statement and the truth, the whole truth, and nothing but the truth. Abraham hadn't told the truth, the whole truth, and nothing but the truth. "Sister" was not the word that characterized his relationship with Sarah. It would have been much more honest to say, "She's my wife. She is also my half-sister." He didn't do that. He said, "She is my sister. Of course, she also happens to be my wife." That was a lie because it was misleading. It was not whole, complete, and simple. A half-truth is a complete lie.

Generally, the truth is simpler than deceit. **The more of the truth I tell, the less explanation I will need.** What God wants from me is integrity of heart and innocence of hands.

◇◇◇

DAY 8 — GOD'S PLANS OF PROVISION

Genesis 22:8 And Abraham said, My son, God will provide himself a lamb for a burnt offering: so they went both of them together.

I usually know when one of my children has a test coming up because of the way he spends his afternoon. He may be a little bit worried or preoccupied about the test because he wants to do well. You know, the hardest tests in life are not the tests for which you prepare just a week ahead of time. The most important tests come in the form of the events of life. They are tests for which you prepare every day, all day, in your normal life.

Genesis 22:2 tells us about a test that God gave to Abraham. He told Abraham to offer his only son, Isaac, as a burnt offering.

What follows is amazing. As Abraham is trudging to the place where he intends to sacrifice Isaac, Isaac asks his father, "Father, you've got a fire, a knife, and wood, but where is the lamb for the burnt offering?" In verse 8 Abraham says, *"My son, God will provide himself a lamb."* That is an extremely powerful and telling verse. It means more than just the lamb that God would provide that day, as you well know.

What did God want from Abraham? Did He want his son? Did He want money, livestock, or some of Abraham's many servants? No, God wanted Abraham. **A person who has given self has nothing more to give and nothing more to withhold. God does not want what you have; God wants you. Here is the wonderful truth: whenever God asks something of you, He has already planned for your provision.**

In the end, God did not take Abraham's son. God gave His own Son. God gave Abraham a promise, and Abraham offered God a son. Abraham's obedience was an act of faith.

Something else indicates this idea of God's planning our provision. Genesis 22:20 says, *"And it came to pass after these things, that it was told Abraham, saying, Behold, Milcah, she hath also born children unto thy brother Nahor."* You can't help wondering why this is included in this chapter. Then you get to verse 23 which says, "Bethuel beget Rebekah." Rebekah would become Isaac's wife. Even as Abraham plotted Isaac's death, God was planning for Isaac's future.

God's Son came through Abraham's son. Abraham's family was part of God's plan. There is a sense in which Abraham had literally tied his destiny to the omnipotence of God. You don't need to know God's grand design to know that there is one. Trust, obey, and tap into God's plan. Remember that whenever God asks something of you, He has already planned for your provision.

DAY 9

PRICELESS OR PROFANE?

Genesis 25:34 Then Jacob gave Esau bread and pottage of lentils; and he did eat and drink, and rose up, and went his way: thus Esau despised his birthright.

In one word, how would you describe Esau as a man? Impulsive. Short-sighted. Immature. Foolish. His life is largely defined by his decision in Genesis 25 to despise his birthright. "Don't sacrifice the permanent on the altar of the immediate." Evangelist Bob Jones Sr. **It is foolish to sell what you cannot afford to lose!**

Hebrews 12:16 warns us by using Esau as an example. The Bible says, *"Lest there be any ... profane person, as Esau, who for one morsel of meat sold his birthright."* The word *profane* means common (as opposed to *holy*, set apart). In other words, Esau treated with contempt ("despised") that which was very holy. **So many times we are just as short-sighted as Esau was**, focusing on what we want right now and forfeiting what we could have in the future.

We ought not treat as common what God has entrusted to us. With a short-sighted perspective, things are common right now that will be priceless in the future. My children come to mind as an example. The older I get, the more valuable the direction that I give my kids has become! I have been thinking now more than ever about college because I have one teenager and two more approaching quickly. If I minimize or fail to even think about this decision that is several years down the road, that is short-sighted—that is "profane." My child's college choice is priceless in the future, so I must consider it important now.

Some of the most valuable things are some of the most unnoticeable, simple things right now. Time, people, and talents are just a few examples. The lesson of Esau is that he was profane—he treated lightly that with which God had entrusted to him. Don't take what you have been given for granted! Something that may not be worth more than a bowl of soup to you now might, in time, be as valuable as diamonds. **Value what you have, while you have it!**

I HAVE ENOUGH

DAY 10

Genesis 33:11 Take, I pray thee, my blessing that is brought to thee; because God hath dealt graciously with me, and because I have enough. And he urged him, and he took it.

Have you ever seen the following scenario play out? It is Christmas time. Two siblings have presents to open. One child gets the super-duper play toy that he has always wanted. His brother automatically wants that *exact* present! If one child gets a different version or accessory, both kids want it. There can be a definite competitive spirit for what someone else has. The great thing about this is, you outgrow such foolishness as an adult, right? Wrong!

Jacob's life was filled with scheming, conniving, and greed. You would never expect to hear the words "I have enough" from him. Yet when he went out to meet his brother, Esau, he was a changed man. Jacob began to take God personally and found that **when you have God, you have enough.** When you don't cling to God and instead do things your own way, you will never have enough!

We think that we lack "things" like money, security, and direction. You do not lack something; you lack Someone. **The question is not how much you have; the question is how much of you God has.** The Lord made this point in Matthew 6:33 when He said, *"But seek ye first the kingdom of God, and his righteousness; and all these things shall be added unto you."* Most of our day is spent getting things, but you will always find something else you need. Your one need today is a Person, the Lord Jesus Christ.

When you have God, you have enough. The alternative is to go after all of the things you think you need, but you will never have enough. Pursue the God Who is all the things that you need today. Can you say this morning, "I have enough"?

DAY 11

DISHONESTY SEES DISHONESTY

Genesis 37:33 And he knew it, and said, It is my son's coat; an evil beast hath devoured him; Joseph is without doubt rent in pieces.

None of us has a perfect home. Now I am not saying that you can't have a godly home and family. I am simply saying that all of us have a horse thief somewhere back in our ancestry. None of us has a perfect lineage because we are sinners. That is certainly true as you read the story of Abraham and his family. You have incident after incident of people who did things wrong and suffered the consequences.

One example is the story of Joseph and his brothers. Joseph's brothers hated him because they were upset with their dad who showed Joseph favoritism. In Genesis 37 they threw Joseph in a pit, sold him to a bunch of slave traders, and basically wiped their hands of any guilt or responsibility for their brother Joseph.

Then, they dipped Joseph's coat in the blood of a goat, brought it straight to their father, and asked him if he recognized it. Verse 33 says that Jacob said, *"It is my son's coat; an evil beast hath devoured him; Joseph is without doubt rent in pieces."* Jacob's sons lied to their father by implication. They didn't state anything. They basically raised a question and caused Jacob to have no doubt about what had happened.

The problem is that what he knew wasn't so. The best liars don't need to state anything. They simply ask a question or leave unsaid something that leads to a falsehood.

Why did Jacob's sons lie to their father? Jacob was deceived at least as much as he practiced deceit. If we go back to Genesis 31, we find that Laban was dishonest with Jacob by changing his wages ten times. In chapter 29, Laban deceived Jacob by marrying him to Leah instead of Rachel. Why would Laban deceive Jacob?

Jacob reaped what he had sown. **A person who is not honest cannot trust anyone.** It is very easy to project our motives onto

16   FIRST LIGHT

other people. A person who lies and deceives cannot trust anyone because he knows that he cannot be trusted.

Secondly, and more importantly, a person who deceives cannot trust anyone because what goes around comes around. We live in a world in which we will reap what we sow. May God help us to plant honesty, integrity, decency, and graciousness to those around us because we will reap what we sow.

HOW WAS YOUR WEEK?

DAY 12

Genesis 41:52 And the name of the second called he Ephraim: For God hath caused me to be fruitful in the land of my affliction.

"How has your week been?" Has anyone asked you that before? The most common answer is "good," but most people don't really mean it! The truth is, **you don't know completely how your day or your week has been**. Remember Joseph? One day he is highly favored of his father, the next he is in a pit. One day he is rescued and sold as a slave to a very rich man, the next he is in prison because of that man's wife. However, God gives him favor, and he winds up second-in-command under Pharaoh.

Joseph summarizes his life—the good and the bad—in Genesis 41. *"God made me to be fruitful in the land of affliction."* Why was this true? Because Joseph was doing what was right. If your standard for a good day is how good the weather is, how people like you, or whether anything bad has happened, your answer most likely is different from Joseph's. A "nice" day is not the same as a day with significance.

Joseph's life was significant and "good" based on his determining to do right; this resulted in, first of all, God's perfect timing. Genesis 41 gives the account of the butler forgetting his promise to Joseph and then later remembering him. But that was God's timing. In other words, *Joseph could have confidence in God's timing because he had determined to do right*.

Joseph's determining to do right also **resulted in God's enabling**. Joseph told Pharaoh as much in Genesis 41:16 when he said, "It is not in me: God shall give Pharaoh an answer of peace." So Joseph's estimation of his life was "good" because of God's enabling and God's timing.

So how has my week been? Ask me five years from now. Your estimation of a day or week might just be amended by history. **History has a way of showing us God's footprint**. "God has made me to be fruitful." That was Joseph's standard of success. He considered **God's will**, not his own; **God's timing**, not his own; and he did it all with **God's enabling**, not his own.

◇◇◇

DAY 13 — PASSING ON A LOVE FOR GOD

Genesis 48:21 And Israel said unto Joseph, Behold, I die: but God shall be with you, and bring you again unto the land of your fathers.

Have you ever thought about what you will leave your kids? You may not have much to leave, but **you leave nothing if everything you leave is something they can put their hands on.** Ironically, inheritances that are left today do not bring the happiness and security for which are intended; often they do more harm and bring misery and heartache. I am not advocating that you leave nothing to your kids; what I am saying is **if you leave only the material, you have done them a disservice.**

In Genesis 48, Jacob was passing on an inheritance. *What* he was passing on was really a *Who*. Jacob was passing on his love for God! Like Abraham and Isaac before him, he did not just pass along possessions; he passed on that love of God. God was passed on to him, and now Jacob was passing this love for God

18   FIRST LIGHT

on to his son and grandsons. Jacob would not be around, but God Almighty would be—*"But God shall be with you."*

What will you pass on when you pass on? Better yet, what *are* you passing on in preparation for your passing on? My wife and I do not have great wealth to pass on to our kids, but we can and should pass on what we know about God and His Word. Whatever material possessions we would pass on will be worthless if we do not pass on the most important "thing"—love for God.

It is not wrong to plan and prepare an inheritance to leave your children. I think it is wise and right. But do not think for a moment that *what* you will pass on can ever replace *Who* you pass on. We are sinners by nature, and giving love for God as an inheritance is never a coincidence. Passing love for God on requires you to be relentless and active, every day and by design. You ought to pass your love for God along to your kids, and with that goal comes purpose and direction with your home, marriage, and children right now. **Whatever you might pass along, make sure you pass along a love for God!**

◇◇◇

WHAT ARE YOUR QUESTIONS?

DAY 14

Exodus 3:14 And God said unto Moses, I AM THAT I AM: and he said, Thus shalt thou say unto the children of Israel, I AM hath sent me unto you.

What would you ask God if you could ask Him three questions today and receive straight, clear, immediate answers? The truth is, you would probably ask the wrong questions. You are better off to let God answer the questions you *should* have today—and He already has.

Moses asks a good question in Exodus 3:11. *"Who am I, that I should…."* Stated in a declaration, Moses says, "I cannot do this, God." And that was true; Moses had no power, no influence, and no way to execute God's plan. Previously, Moses had seen two

Jewish men fighting, and he could not even get those two men from his own country to follow him, much less a whole nation.

But God never answers Moses' question; He does answer the question that Moses should have asked. In verse 12, God says, *"Certainly I will… ,"* and in verse 14, He says, *"I AM."* Whatever may be your "am I" question, God's answer is "I AM." You may wonder today, "How am I _____?" God's answer is, "I AM." "What am I to do about _____?" "I AM." "When am I _____? ""I AM."

It is not about us; it is about God working through people like us. "Who am I?" or "Look at me" will both lead you astray. It is not as though you will impress God with what you have—He gave you everything you have.

For everything you are not, and for everything you need, God says, "I AM." That means **He is the Eternally Existent and All-Sufficient One, the Creator of the universe. And you can trust His answer to your questions today!**

◇◇◇

DAY 15: A HARD HEART EQUALS A HARD LIFE

Exodus 8:15 But when Pharaoh saw that there was respite, he hardened his heart, and hearkened not unto them; as the LORD had said.

When I was growing up here on the Ranch, my mom always had a summer garden. Now, Mom did all the planting, but her children did much of the cultivating. For fifteen minutes every morning we would weed the garden, which in Middle Tennessee means pulling out rocks. I was always amazed that I could spend fifteen minutes pulling out rocks one day, and the next day there would be rocks all over again. No one in his right mind cultivates rocks for a garden if he is trying to grow something edible. Yet there is something even harder than rocks that many of us are cultivating today, and that is a hard heart.

Pharaoh is an example of a person who had a hard heart. God began to judge Pharaoh because of his rebellion. Exodus 8:15 says, *"But when Pharaoh saw that there was respite [relief from the plagues], he hardened his heart, and hearkened not unto them."* A hard heart often manifests itself in a person who refuses to hear. It is important for us to know that **a person who cultivates a hard heart will reap a hard life full of needless pain and hardship.**

What did Pharaoh cultivate that culminated in a hardness of heart? First of all, he cultivated pride. In Exodus 5:2, Pharaoh says, *"Who is the Lord?"* When I refuse to acknowledge God, I am going to be unable to know Who He is and understand life. Pride is the native sin in each of us and brings about much misery in our lives.

Pharaoh also cultivated procrastination. In Verse 10, Moses says, "Hey, would you like me to call off the plagues to entreat the LORD for you?" Pharaoh says, "Yes." Then Moses asks, "When would you like me to do this?" Pharaoh famously says, "Tomorrow." That is in stark contrast to verse 23 where God put off a plague until the next day. God was putting off the plague, and Pharaoh was putting off the relief from the plague. If I decide when and how I obey, I am not obeying.

In verses 25-28 Pharaoh took the initiative and told Moses and Aaron, "Go ye, sacrifice to your God in the land." Moses responds, "No, that is not the way God wants us to do it. We will go three days' journey into the wilderness and sacrifice to God as He commands us." The reason that Pharaoh took initiative in responding to God was that he felt like he could draw up the terms of surrender.

You negotiate with someone you feel to be your equal. Pharaoh took the initiative with God but with his own plan, as if he and God were equals. God doesn't want us to negotiate, procrastinate, or act in pride. God wants us to hear His voice, obey His commands, and reap the life of peace that comes.

DAY 16
DON'T LET GOD'S MERCY BE YOUR JUDGMENT

Exodus 9:4 And the LORD shall sever between the cattle of Israel and the cattle of Egypt: and there shall nothing die of all that is the children's of Israel.

Pharaoh was ignorant of God because he was arrogant against God. What he did not know about God was based on what he thought of himself. In Exodus 5:2 Pharaoh rather arrogantly says, *"Who is the LORD, that I should obey his voice?"* Much of the rest of the next few chapters is an answer to that question from this arrogant king.

In Exodus 9 God sent plagues on Egypt. Each plague and answered to an individual Egyptian God. They were attacks against the gods that Pharaoh and Egypt had trusted. Exodus 9:1 makes, *"The LORD God of the Hebrews."* Who is this God who is talking to Pharaoh? It is the Lord God of the Hebrews, the God of the people of Israel.

So, God answered it in context and God answered in judgment. In this chapter we read about a plague that he sent upon all the animals of the Egyptians, the horrible boils that infected these people of Egypt, and a hail that was mixed with fire. The book of Exodus says that God did all this so that the Egyptians would know Him to be the all-powerful God of the Hebrews.

Who is the LORD God? When you read these chapters, you come to the conclusion that God is a God of power, judgment, and anger. Is He merely a God of anger? Is He merely a God of mercy and grace? So many times we make our own god by constructing Jehovah as we wish Him to be. God is Who He is; and God reveals Himself to be Who He is, in some measure, depending on how we live before Him.

Was God showing Israel mercy, or was God showing Egypt judgment in Exodus 9? The answer is "yes." God's judgment on Egypt was God's mercy on Israel. There is a very important lesson here. **Never put yourself in such a position that God**

must give judgment to you in order to give mercy to another. God was giving mercy to Israel by giving judgment on this implacable, intractable, arrogant, stubborn Pharaoh.

Wherever you are, remember that God is Who He is. Make sure that you receive mercy from God by showing mercy to others. You can know God's mercy today if you will show mercy, and you can know God's grace if you will be humble.

GOD WILL WIN

DAY 17

Exodus 9:16 And in very deed for this cause have I raised thee up, for to show in thee my power; and that my name may be declared throughout all the earth.

On June 27, 1876, well over 100 years ago, General George Armstrong Custer led his men straight into defeat. What Custer thought would be an easy victory against a village actually turned out to be total destruction for him and his men at the hands of the Sitting Bull and Sioux nation. I do not like General Custer, nor is he one of my heroes. But although the Sioux won the Battle of Little Bighorn, they lost the war. It was bound to happen. You see, Custer was defeated, but there were a lot more where he came from.

I do not care what "the score" may be right now; **the truth is, God will win the war.** He always does. But *how* He wins is up to you. God can work through you or against you; with you or without you; in this lifetime or in eternity. Whether you see it or you do not, God will win.

With God, there is no such thing as a "win-win" or a "win-lose" situation. One word describes God's record—WIN. **God will win.** He is the One Who raised up Pharaoh (Exodus 9:16). Do you suppose Pharaoh realized this? Did the Jews know this?

God raised up Pharaoh for His (God's) glory and to show His (God's) power.

Throughout Exodus 8-9, God makes known that He will win. Exodus 8:10 declares, *"That thou mayest know that there is none like unto the LORD our God."* Verse 22 repeats, *"To the end thou mayest know that I am the LORD in the midst of the earth."* Chapter 9, verses 14 and 29 say, *"That thou mayest know that there is none like me in all the earth,"* and, *"That thou mayest know how that the earth is the LORD'S."*

We should be neither discouraged in doing right nor arrogant in doing wrong. God will win. He will win today, and you can ask Pharaoh about that.

◇◇◇

DAY 18

REMEMBER

Exodus 13:14 And it shall be when thy son asketh thee in time to come, saying, What is this? that thou shalt say unto him, By strength of hand, the LORD brought us out from Egypt, from the house of bondage.

Like elephants, kids never forget. They always seem to remember when parents promise ice cream or when parents recall their most embarrassing moments. **But your kids will only remember what you reveal to them.** They will not have an awareness of God's presence in their lives if their parents do not cultivate it.

You ought to show them God's hand in your family history. Exodus 13:3-16 is about a memorial the Israelites erected in order to teach their kids about what the Lord had done in their past. Verse 8 says that parents were to purposefully teach their kids that *"this is done because of that which the LORD did unto me when I came forth out of Egypt."* How many people came out of Egypt? Was it not thousands? And yet dads and moms were to teach their kids what the Lord had done for them personally. The same is true for you. **Your kids should see God's**

working—not just in a corporate, generic sense—but in a personal way in your family.

I come from a family of storytellers. Both sides of my family are good at telling stories about our family's past. It is easy now to look back one hundred years through the stories that have been passed down. But what if you are the first one to know God in your family? That does not change the need for you to rehearse to your children God's dealings with your family throughout history. Just like the children of Israel, you can teach your family what you were, what you are now, and that God is the reason why it is so.

Parents, it is up to us to keep alive, to illustrate, God's working to our children. Answers to prayer and God's leading in our lives are two great areas to remember. It takes real work to let kids know when an answer comes to what we have been praying for. Often we can pass right over the fact that God has worked in our life and in our family. If answers come, but we do not teach our kids, they will never know.

Make good memory a goal in your family. Take the responsibility personally and see to it that your kids see and know God's working.

DAY 19
IS HIS PRESENCE PRESENT?

Exodus 33:15 And he said unto him, If thy presence go not with me, carry us not up hence.

Have you ever seen a person at the store who obviously stood out? Your first reaction might be, "Wow! That's… different." Some people are just so unique in their walk, talk, or dress that you couldn't possibly miss them!

We often associate "unique" with weirdness; but the truth is, we should stand out as different from the world around us. **Nothing distinguishes you better than the presence of God.**

A year ago, I was at a store in Arizona, and I saw a man dressed in obvious and distinct religious clothing. You could tell by his clothing that he was different than most of the people in that town. Truthfully, I was different than most everyone else in that city, too. I was even more distinct in God's sight because of Christ.

No way is good that is not God's way, and no place is good that does not have God's presence. When you look at the future, there are all kinds of things about which you could worry. What determines the future is not *where* you are going but *Who* is going with you.

Moses' prayer was not that God would send the children of Israel, but that He would lead them. Moses believed that any place was a good place if God was there! Nothing would distinguish God's people more than God's presence.

It is easy to look at other Christians and think that they are important because of the money, the influence, or the ministry they have. However, the most important thing in life is God's presence. We distinguish ourselves from the world by having the presence of God. As Moses said, *"If thy presence go not with me, carry us not up hence."*

SKILL AND WILL

DAY 20

Exodus 35:5,10 Take ye from among you an offering unto the LORD: whosoever is of a willing heart, let him bring it, an offering of the LORD ... And every wisehearted among you shall come, and make all that the LORD hath commanded.

Is it more important to have skill or willingness? Well, that may be a misleading question! **Skill and willingness are not mutually exclusive.** It is best to be willing in what you receive and skilled in what you do. There is no reason to not be both!

In Exodus 35, God was looking for willing hearts and skilled ("wise") hearts. Both are necessary. I have said, and have heard said many times, that **God is not looking for skilled people; God is looking for willing people.** That is true. You can hone the skills you have been given by God, but willingness is a decision of your will. A willing heart must come before a skilled heart, but both are important!

"I can" and "I will" are two different things. Sometimes we say, "I can't," when in reality, we mean, "I won't." God's work is not left for those who do not have the skills to do anything else. However, it does not matter if you can do everything if you are not willing!

It is easier to teach a willing person skill than it is to teach a skilled person willingness. You need willingness, and you need skill. Some of the best summer camp counselors we have at the Ranch are young people who are willing. It turns out that the willing ones are pretty skilled and are willing to learn, too!

My dad once heard his uncle John Rice say, "I am a man void of very many gifts, save the ability for hard work." Now, I think it is obvious that John Rice was a highly skilled man, but his point was that anything God had accomplished through his life was based not on his skill but on his willingness. Work hard. Work well. Be willing as you use your skills today.

DAY 21

JUST DO IT

Exodus 36:1 Then wrought Bezaleel and Aholiab, and every wisehearted man, in whom the LORD put wisdom and understanding to know how to work all manner of work for the service of the sanctuary, according to all that the LORD had commanded.

When you do what you should do for the day, you will have what you need for the day. God's plan is much bigger than your capacity to know it! God is a God of detail, and He knows the "big picture."

I am sure that Bezaleel did not know every detail of the work God had for him. Exodus 36:1-2 says of him, *"Then wrought Bezaleel… the work to do it."* There came a time to get the work done! They needed to "just do it."

God's plan is far bigger than your part, but it includes your part. You may not know exactly how your part fits into His plan, but that does not make your part unimportant. Someone had to give the goats' hair in order for someone else to make curtains with it. When we are listening to and following God, there is confidence in the plan and provision God has!

I don't know what your part is in God's plan; but I do know that **if you do what you should, you will have what you need.** When you take the step you should take now, God will reveal the step you should take tomorrow. The Master has a master plan, and your partnership in His plan is important. Most of your questions are probably about details, but when God has clearly given you a part—in the words of a famous shoe maker—*just do it*!

IGNORANCE IS NOT BLISS

DAY 22

Leviticus 4:2 Speak unto the children of Israel, saying, If a soul shall sin through ignorance against any of the commandments of the LORD concerning things which ought not to be done, and shall do against any of them:

As I crested a big hill on Interstate 70, I saw what no one wants to see on the left side of the road: a police officer. Immediately I did what most of you would do—I pressed on my brake pedal to slow down. Even though I was simply passing a slower driver, do you think the officer would have understood and apologized for pulling me over? **A law that is broken is still broken, regardless of the excuse—even if the excuse is ignorance.** For instance, my youngest son Weston sometimes claims, when he breaks a "law" in the Rice house, "Daddy, I forgot!" His punishment will help his memory, but **forgetting is not an excuse for doing wrong.**

The sin offerings detailed in Leviticus 4 are instructive about our life. While you may not be Jewish, you <u>are</u> part of Abraham's family by faith in Jesus Christ. While you and I do not sacrifice animals according to the measures described here, we can learn two lessons about our lives. First, **we are accountable for what we do.** The rest of this chapter covers anyone and anything. From the "whole congregation" (verse 13) to a "ruler" (verse 22) to the "common people" (verse 27), no one is above the law or below the law. Notice the repetition of the word "guilty" in each of those verses; that gives a sense of responsibility. Isn't it amazing how many people from all walks of life think they are above the law today? Over and over again, you hear of high-ranking executives of financial institutions who think they are above the law. But God holds us accountable for what we do.

Not only are we accountable for what we do, **we are even more accountable for what we know.** Verse 14 says, *"When the sin, which they have sinned against it, is known, then…."* Verses 23 and 28 say, *"Or if his sin … come to his knowledge…."* So then, **God holds us accountable for what we know and when we know it.** Would these Jewish people have been better off to sin in ignorance? Would ignorance be acceptable instead of sacrificing all the animals? The answer is, "No"—ignorance is not bliss. We are not just responsible for what we do; we are even more responsible for what we know.

No one enjoys a pricked conscience. While the feeling may not be exciting, it is not a bad thing to be reminded to do the right thing. Enjoying life after today requires doing what is right today. And may that be the case for us today!

◇◇◇

DAY 23
ARE YOU CONTAGIOUS?

Leviticus 13:4 If the bright spot be white in the skin of his flesh, and in sight be not deeper than the skin, and the hair thereof be not turned white; then the priest shall shut up him that hath the plague seven days.

Have you noticed how conscientious churches are these days about good hygiene? That is not bad; that is good! More and more churches have a bottle of hand sanitizer in the lobby. Some people in the congregation will forego shaking hands when they feel they may be contagious.

I must admit, Leviticus 13 is a chapter that makes you want to wash your hands after you read it! The whole chapter is about contagious diseases like leprosy. It occurred to me that **if we were as concerned about our moral hygiene as we are about our physical hygiene, we would be much better off as Christians.**

The truth is, everybody is contagious. Our conditions are contagious. No one is unimportant. No man lives to himself or dies to himself.

The only thing more contagious than a cold or the flu is attitude. The only thing more contagious than spreading germs on your hands is spreading sin in your heart. **Whatever we allow to spread in us will spread to those around us.**

Today, hopefully you will wash your hands several times. As you do, take a moment to think about what else you might be spreading to others. Make it a point to keep only what is worth spreading!

◇◇◇

DOES GOD CARE ABOUT CLEAN?

DAY 24

Leviticus 14:54-57 This is the law for all manner of plague of leprosy, and scall, And for the leprosy of a garment, and of a house, And for a rising, and for a scab, and for a bright spot: To teach when it is unclean, and when it is clean: this is the law of leprosy.

Just the other day, for most of the day, I cleaned out the barn behind my house. My barn is the place where—perhaps like you—I accumulate all kinds of junk. I found sink parts (left over from a previous repair), screen for a screen door (in case there is a future repair), and various other things I've been stashing away since the last time I cleaned out the barn. I have learned that the security of keeping parts and seldom-used items has to be balanced with how likely it is that I will ever need them! That is why I spent the good part of a day "cleaning house."

Leviticus 14 deals with how to take care of contagious things. **God not only wanted the physical to be clean and holy, but He also wanted the lives of His people to reflect that.** You probably will not find an in-depth study of chapter 14 in Sunday School this Sunday, but these would be important matters if you lived in a community of several thousand people. Some Bible teachers believe the "leprosy of a house" was some type of mold that could infect your physical house. If living in close quarters, you would be interested both in knowing how to take care of the mold in your house or the disease in your body, which was highly contagious.

Knowing how to take care of contagious things not only applies to cleaning our houses but also to cleaning our lives. It is important to actively clean the mold, clutter, dirt, and grime in your house today; in the same way, you ought consider how clean your life is today. **Cleaning is important because things build up—they can accumulate.** Just like dirt in your house, "dirt" of the world tends to build up also. You ought to check your life and clean out any dirtiness you find.

Cleaning is also important because things spread and can be contagious. Wouldn't you want to know if your neighbor had a contagious disease? You would want to stay away if you could help it! Often people will not shake hands at church if they have a cold; they don't want to pass along the germs. But not even a cold can spread as quickly as a bad attitude. Someone's rotten, complaining attitude can spread like a contagious disease in a matter of minutes! That is why you must clean your life today—because things spread.

Reading Leviticus will remind you of dirt, grime, and contamination. **Let your thinking about hygiene remind you to think about your life. Purity is important to God—today is a great day to do some cleaning!**

◇◇◇

DAY 25
WHAT'S A NAME WORTH?

Leviticus 22:32 Neither shall ye profane my holy name; but I will be hallowed among the children of Israel: I am the LORD which hallow you.

What is a name worth? Well, it depends on whose name it is, but I can tell you that many people have become wealthy by having the right signature on the right line! Some signatures are even more valuable years after the person is dead, if he or she has some historical significance.

A name is not valued by its monetary gains alone. Money can actually be the poorest of measurements for a name! The truth is, some names are important because of their integrity and authority. With the right name on a particular document, you can have credibility and authority you would not have otherwise.

No name is worthy of more honor than the name of God Almighty. Live today in such a way that a holy God could sign His name to it. The Bible tells us to *"do all to the glory of God."*

Would God be able to sign His name, to give authority and power, to your actions?

The name of God (LORD) tells us that **He is the eternally existent One.** He is the Creator. Pause and take time today to consider if the high, holy, self-existent God could sign His name to your actions. Profaning God's name doesn't just include cursing or swearing; profaning God's name means to treat His name flippantly or tritely. **You can profane His name by your words *and* by your actions!**

The way to properly consider His name is to follow His Word. Psalm 138:2 tells us that God has *"magnified [His] word above all [His] name."* What an amazing thing! There is no manifestation of God's essence more magnified than God's Word. **What He has said reveals Who He is.**

How will you live today? Will you live dominated by the urgent? By the whims of others? By your own passions? Or will you live a life today that God would be pleased to sign His name to?

◇◇◇

HE REALLY OWNS IT ALL — DAY 26

*Leviticus 25:17 **Ye shall not therefore oppress one another; but thou shalt fear thy God: for I am the LORD your God.***

Here in Leviticus 25, the Bible describes the Year of Jubilee. Every fifty years, the land lay dormant, servants were freed, and any land that was purchased was returned to the original owner. If a Jewish man came upon hard times, he could sell himself and sell his land. But at the Year of Jubilee, he received back his land and his freedom.

We can learn from this chapter that **when you claim God, He claims everything else.** The ownership of God was based on two things: God owns the land, and He owns the people on the land. In dealing with servants and land, the Lord did not want them to take advantage of one another. He says in verse 23, *"For the land is mine; for ye are strangers and sojourners with* me." Verses 38 and 42 say, *"I am the LORD your God … they are my servants, which I brought forth out of the land of Egypt."* Everything you do today should be based on the fact that you are a steward of God's things.

Suppose I sat down and figured out how much money I could make by selling things here on the Ranch. The "stuff" here at the Ranch—houses, pickup trucks, equipment—might have some monetary value. But the truth is, none of it belongs to me—in a very real sense (Bill Rice Ranch, Inc. owns it) and in an ultimate sense (God owns it!). In dealing with the details of loaning money, buying land, and buying servants in light of the Year of Jubilee, God's foremost concern was, *"Ye shall not therefore oppress one another…."* The children of Israel were not to take advantage of others in loaning money (verse 37) because their attitude was to be based, not on what the debtor could do in return, but on what belonged to God.

That same attitude should make a difference in our lives today—God owns us, our car, our furniture, our kids, etc. **Understand that God owns everything, and you simply have a stewardship.** Although the Year of Jubilee was a distinctively Jewish event, no matter how little or how much you own, **handle life today with the same realization that when you claim God, He owns everything else.**

◇◇◇

DAY 27
DO NUMBERS MATTER?

Numbers 1:2 Take ye the sum of all the congregation of the children of Israel, after their families, by the house of their fathers, with the number of their names, every male by their polls;

Do you know what the temperature was this morning? Do you know how old you are? How many inches are in one foot? "Brother Wil," you interject, "what do these questions have to do with *anything*?" Well, do numbers matter? Of course they do! You probably do not value what you do not count.

Let's say (hypothetically, of course!) that you have a "chat" with a nice officer along the side of the road about your car's speed. Two numbers matter: the number on the speed limit sign and the number to which your speedometer needle was pointing! Numbers are important!

You cannot steward what you have if you do not count what you have. The book of Numbers is all about numbers. God told Moses to take stock of the number of men able to go out to war. To properly fight in battle, you have to know the size of your army and the weapons at your disposal. There is no way to prepare if you do not number.

There is probably something each of us counts every day. You may count your steps, your bank account, your mutual funds, or the students in your class, but we all count something. **You will number what is important and valuable to you.**

So what numbers matter to *you*? What are you actually counting? You will count whatever you see as valuable. God cares about numbers because God cares about people. May He help us to be good stewards of what He has given to us!

BEARING BURDENS

DAY 28

Numbers 4:15b After that, the sons of Kohath shall come to bear it: but they shall not touch any holy thing, lest they die. These things are the burden of the sons of Kohath in the tabernacle of the congregation.

The rebellion of Korah is well-known and documented for us in Numbers 16. To fully understand Korah's rebellion and God's judgment, you must go back to chapter 4, where the Bible says, *"These things are the burden of the sons of Kohath...."* Korah was a "son of Kohath," so his God-given responsibility was to be the "moving van" for the most sacred parts of the tabernacle. In Korah's mind, what God had put in his charge was not big enough! (16:9)

Korah became so busy with others' "burdens" that he would not take care of the burden he was entrusted with by God. Don't bear the burden of the world on your shoulders. **Simply lift the burden that God gives you.** What God has given you is more than enough!

Sometimes I read about decisions people, primarily in leadership, had to make. After reading about some of the decisions the President of the United States has to make, I wonder why in the world people are eager to run for election! It is easy to think, "I could do better than that!" Perhaps at times I could do better, but that is not my burden to bear. God has not given me the weight of a country, but He has given me a specific burden to bear. Lift the burden God gives *you*, not the burden He gives someone else!

Don't let your days be filled with worry and frustration about someone else's burden. Just bear the burden God gives *you*! It may not always be wrong to seek another position or a new responsibility at work or at church. The point is, **do not live your life weighed down by a burden that God never gave you.** If you will carry the burden God has given you, it will help everyone else carry their own burdens!

Moses, Aaron, and Korah were all just servants with different God-given responsibilities. Korah got into trouble when he became obsessed with bearing someone else's burden. We should all be shooting for the same goal, but we all have individual responsibilities. Are you bearing a burden today that God did not give you? Thank God for the burden He has entrusted to you, and **do your best to serve Him and bear that burden today**!

WHAT DO YOU SEE?

DAY 29

Numbers 13:30 And Caleb stilled the people before Moses, and said, Let us go up at once, and possess it; for we are well able to overcome it.

Do you remember the kids' song about the twelve spies that said, "Ten were bad and two were good"? Numbers 13 records those spies' reports of the Promised Land. What do you think they saw? Did the two "good" spies see something different than the ten "bad" spies? Both groups saw a fruitful land that flowed *"with milk and honey"* and was filled with grapes, pomegranates, and figs; but they also saw the giants, strong fighters, and walled cities. According to each of their reports, **they saw the same things, but they did not see the same things!**

Their differing reports proved that **what you *know* determines what you *see*.** All twelve spies saw the giants, but only Caleb (and Joshua) said, *"We are well able to overcome it."* What did Caleb and Joshua *know* that helped them *see* what they saw? What experience did they have? Well, they had experienced God's hand in parting the Red Sea and defeating the mighty army of Egypt. Plus, the land they were spying was called the *Promised* Land—promised by God! That is what Caleb and Joshua knew, and it affected what they saw!

The facts about the Promised Land were not disputed. While all the spies had the facts, **only two had *faith*.** In your life today, what promise are you standing on? What you see today will be determined by what you know. If you have the confidence in knowing that you are doing what God wants you to do, that will affect what you see today. Do you see only the bad—the giants, strong armies, and walled cities? Or do you see those same things in light of the promises of God? You cannot wish away your circumstances, but you can live your life based on God's promises and power.

God has not promised us a life without "giants" or difficulty, but He has given us *"great and precious promises"* in His Word. When you stand on what God has promised, all the negatives in your life will be put into proper perspective. What do you see today? Well, that will depend on what you know!

◇◇◇

DAY 30
NOTHING BUT THE WHOLE TRUTH

Numbers 16:3 And they gathered themselves together against Moses and against Aaron, and said unto them, Ye take too much upon you, seeing all the congregation are holy, every one of them, and the LORD is among them: wherefore then lift ye up yourselves above the congregation of the LORD?

When a witness is sworn in by a court of law, he promises to tell what? "The truth, the whole truth, and nothing but the truth." Is there really a difference between the truth and the *whole* truth? Between the truth and *nothing but* the truth? In a sense, oaths are made for liars! There is a difference in **the truth, the whole truth, and nothing but the truth**; and Numbers 16 illustrates the difference.

Korah was a rebel, and his rebellion spread to 252 other people, a rebellion which ultimately resulted in the death of nearly fifteen thousand people. There is a lesson to be learned about rebellion, leadership, and influence; but I want us to look at the statement by Korah and his gang in Numbers 16:3. They said, *"All the congregation are holy, every one of them, and the LORD is among them...."* Were the children of Israel set apart (holy) to God? Yes, they were. Was the LORD among them? I think you could say yes. Was Moses in authority over them? Yes, he was. So was what Korah said true? No! He did not tell the truth, the whole truth, and nothing but the truth!

The question you must ask yourself is, **how much of the truth are you telling?** I have a friend at church that bagged one of the biggest bucks on the Bill Rice Ranch last year. Impressed? Well, you might not be as impressed when I tell you that his Chevy truck did all the damage on Highway 96! It's true that he killed a buck; the whole truth makes the story different altogether! In the age of Twitter, Facebook, and YouTube, it is vital to keep the truth, the whole truth, and nothing but truth in mind, both in what you say and what you hear.

How much of the truth are you believing? It is a dangerous place to be when you only know half of the facts! Knowing that you have the whole truth instead of half the truth will help you in your daily life. If you hear that someone said something about you, be dead sure that you have the whole truth before confronting them. As a parent, be sure you have all the facts before the punishment is given out. Sometimes a sister is good at framing a brother! (It works the other way, as well.)

Be a person of honesty and integrity in what you say, and be careful about jumping to conclusions without knowing all the facts. **Is your life filled with the truth, the whole truth, and nothing but the truth?**

◇◇◇

IS IT WRONG TO WANT?

DAY 31

Numbers 17:5 And it shall come to pass, that the man's rod, whom I shall choose, shall blossom: and I will make to cease from me the murmurings of the children of Israel, whereby they murmur against you.

Korah was discontent with who God had made him and what God had given him. Korah expressed his rebellion to God by showing discontentment to Moses and Aaron. The truth is, **you will always be discontent if you look at other people instead of focusing on what God has given you.** Whenever you look at someone else and make comparisons, you are hiking down a trail that never ends well! You will either end up proud or jealous, but neither one is good.

Do not look at another person as the standard of God's will for you. You can't do any better than accepting what God has given you and who He has made you to be. So then, is it wrong to want? The answer is, No! You wouldn't be alive right now if it wasn't for wanting. It is not wrong to have ambitions, but it is wrong if your wanting is pitted against God's way.

Murmuring and grumbling are essentially expressing your will against God's way. There will always be someone who has more than you do and someone who has less than you do. But there will never be anyone who has what God has perfectly given you!

What should you do when you don't know if what you want is right? Ask God about it! It's much better to ask for what you want than to complain about what you don't have. **Contentment is accepting God's will and way; murmuring is pitting my want against God's will.** It is not wrong to want, but make sure you submit what you want to God's will.

◇◇◇

DAY 32
DISMISSING GOD

Numbers 20:10 And Moses and Aaron gathered the congregation together before the rock, and he said unto them, Hear now, ye rebels; must we fetch you water out of this rock?

Imagine attending a birthday party thrown in your honor; but when you (as the guest of honor) arrive, you find the reception somewhat cold and indifferent. Everyone is excited and talking to each other, but no one talks to you. Everyone brings presents, but they are giving them to each other, not to you. Wait a minute! You are the reason for the party, but you are dismissed and ignored in the corner. That would be bizarre, wouldn't it?

This type of scenario is exactly what played out in Numbers 20. A bunch of rebels complained about the lack of water, and a rebel struck the rock in anger and rebellion. The One who gave water in the first place was disregarded by both!

When you fuss with authorities in your life, you complain against God. That is a dangerous thing to do! On the other hand, when you as an authority respond in rebellion, you have done the very same thing. **Doing God's work in your own fleshly passions is rebellion.**

Both Moses (the rebel) and the children of Israel (the rebels) missed God in this entire account. **You accuse God by complaining, but you dismiss God by responding to people without regard for Him.** Be careful that in your day you don't try to combat wrong by doing wrong. Don't combat rebellion by being a rebel. Don't dismiss God in your actions or your reactions!

NAVIGATING WITH PATIENCE
DAY 33

Numbers 21:4 And they journeyed from mount Hor by the way of the Red sea, to compass the land of Edom: and the soul of the people was much discouraged because of the way.

When I need directions to a location, I normally depend on my iPhone map or GPS to give route options. Most times I am given three options to get to where I want to go.

Sometimes in life, you don't get to choose the route; the route chooses you. Maybe that is where you find yourself this morning. You feel like you are not taking the direct route to where you want to go. The children of Israel found themselves in the same spot after forty years (!) of detours and discouragements. They became impatient because of the way.

If you are on a detour, you have a choice. **Sometimes faith looks a lot like patience.** Clearly, there are times when faith means "go," but there are some times when faith means "wait." Impatience will breed discontentment and poor assumptions in your life.

Maybe you feel like you are on a route that is out of the way. **If you are doing what you should do today, trust God and be patient in the way.** Don't allow discouragement and discontentment to cloud your navigating.

DAY 34
BACKDOOR SINS

Numbers 25:18 For they vex you with their wiles, wherewith they have beguiled you in the matter of Peor, and in the matter of Cozbi, the daughter of a prince....

There is evil in the world. There are people who disdain what is good, wholesome, and right. Today, you are considered immoral by holding to biblical principles about anything. Satan would love nothing more than to destroy you (I Peter 5:8), and he is just as willing to do so with wrong friends as he is with the right enemies.

A man is much more vulnerable to his own vices than to an enemy he hates. **It is much easier to destroy a person with something he likes.** It is easy for us to thunder against the sins we dislike but yet be blindsided by our own passions.

What Balak, king of Moab, could not do in cursing and destroying Israel, Israel did to themselves when they settled at Shittim. Numbers 25:1-2 says that they *"began to commit whoredom with the daughters of Moab, and they called the people unto the sacrifice of their gods...."* An army of enemies couldn't destroy Israel, but the daughters of Moab could by seducing them into wickedness and idolatry.

The truth is, **the devil is a master of blindsiding you with "backdoor" sins.** You are not likely to fall to something you absolutely hate. The thing you will be tempted with will be something which has an ally in your own needy heart!

Don't become a casualty to "backdoor" sins like envy, malice, discontentment, jealousy, and gossip. These types of sins that find an ally in your heart will *"vex you with their wiles."* The devil is not ignorant. He is savvy to what will blindside you today. Be on guard and watch the back door!

DEFEATING GIANTS

DAY 35

Deuteronomy 2:24 Rise ye up, take your journey, and pass over the river Arnon: behold, I have given into thine hand Sihon the Amorite, king of Heshbon, and his land: begin to possess it, and contend with him in battle.

As a child, didn't you enjoy hearing a good story? Maybe you liked a bedtime story or a story about the old days from a grandparent. My great-grandmother, who was a farmer's wife in South Dakota, used to tell me stories about the natives of the land, gypsies, tumbleweeds, and farming. I still remember that!

Whether we know it or not, we all tell ourselves stories every day. We tell stories about who we are, what we can do, what we should not do, what has been, and what will be in the days ahead. Take care as to the kind of stories you tell yourself. **The story you tell yourself about the past will, in large measure, determine the story you will live out today.**

As Israel prepared to enter the Promised Land and stood on the cusp of both great danger and great promise, Moses told them a couple of stories about what had been, precisely because of what was about to be. The land was full of giants (verses 9-23), but God had already given it into Israel's hands. Sure, there were giants, but God was greater than the giants!

There is no stronghold too strong for people who rely upon God's strength to possess their inheritance. There is no giant on the path of God's will that is undefeated. You may see a giant you have never faced before, but there is no giant that has not been faced and defeated before by the power of God. The question is not, "Is this an easy path?" The only question is, "Am I following God?"

You can't always choose your giants, but you choose victory when you choose to follow God. Tell yourself that story today!

DAY 36
DESIGNED FOR VICTORY

Deuteronomy 7:1-2 When the LORD thy God shall bring thee into the land whither thou goest to possess it, and hath cast out many nations before thee ... seven nations greater and mightier than thou; And when the LORD thy God shall deliver them before thee; thou shalt smite them, and utterly destroy them; thou shalt make no covenant with them, nor shew mercy unto them.

Have you ever met a loser? I mean a real-life, sure-enough, all-American, A-1 grade LOSER? Maybe you feel like *you* are the loser. You may be winning in the sight of people; but you have this secret, gnawing fear that it's all a facade. Simply put, you think you are not a winner when it comes to the Christian life and serving God.

I can empathize with the kids' cartoon that says, "I know I'm somebody 'cause God don't make no junk." I believe that is true, but I am not "somebody" because I am great. I am somebody because God is gracious.

The book of Deuteronomy is the story of God giving and Israel possessing. If God is giving it, we ought to have it. Someone wisely said, "Seek no more than God gives, but settle for no less." Moses spoke about the land that God gave and Israel was to possess as if it were a done deal! His instruction to them was clear: "When God gives victory, do not make a place for defeat."

Can I remind you of something this morning? **God designed you for victory.** There is no reason you cannot be everything that God made you and intended you to be. Do you honestly think that God made you for defeat?

I do not mean that you have a perfect body or perfect mind on this earth, but I do mean that your heart should be willing to accomplish all that God intends for you. Since God intends victory, **any defeat is an "inside job."**

God loves you because *He* is good, not because you are. You have God with you today. Don't accept defeat, because He created you for victory. Don't sabotage victory by harboring in your heart that which God will not bless. I don't know what your intentions are for the day, but **God's intention is clear: He wants to give you victory on the path of His choosing.** Accept no less!

DO YOU REMEMBER?

DAY 37

Deuteronomy 8:14 Then thine heart be lifted up, and thou forget the LORD thy God, which brought thee forth out of the land of Egypt, from the house of bondage.

Whenever people face a natural disaster, they are forced to protect their most precious possessions. It is amazing how often those possessions turn out to be photographs of the life they have already lived! Memory is a very precious thing. It can be a blessing, or it can be a curse.

A good memory is crucial to a good relationship with God. Sometimes we forget Who He is and Who He has been. We forget where we are and where we have been. We forget how weak we are and how strong God is. The human tendency is to forget.

Moses said to the children of Israel that God *"humbled thee, and suffered thee to hunger...."* That doesn't sound very nice, does it? God allowed them to hunger. Does that mean that God is not good? No, sir! He allowed all of their problems so that He could be their answer. **The great danger was not starving; the great danger was being full** *and forgetting.*

God wants you to remember what He has done. It is far better to have problems and see God, than to have no problems and not see God. **Pride is mistaking God's provision for our own power.** Don't let your heart be "lifted up." It is not your power; it is God's grace.

Whatever lies before you today, take a moment and look back at Who God is, what He does, and what He is able to do.

DAY 38

PASS IT ON!

Deuteronomy 26:5a And thou shalt speak and say before the LORD thy God….

One thing that ancient civilizations have in common with us is the tools that we are *not* using. They were just beginning to use the tools of pen and paper and books; we are leaving that era! Whatever our tools now, **it is important that we record and remember what God has done.** If we don't, we will make the same mistakes that have already been made.

Deuteronomy 26 shows the oral tradition the children of Israel had since they did not have books. They did not have the book of Deuteronomy—they were *living* Deuteronomy! Without pen and paper, they relied heavily on spoken (oral) tradition. Similarly, we are moving from a literate society to a digital society. You may have actually read more the past three years than you ever have; but your reading is by scanning, and your writing is by typing. Now, instead of reading a book, you can just download the same book in audio format to your device!

Nevertheless, it is just as important today as it was in Old Testament times to remember what God has done. One way you can do that is to **orally remember and to remind others of what God has done.** I don't know if the people in Deuteronomy 26 recited verses 5-10 word for word every time, but on a regular basis they recounted what God had done! It is very important to God that what we know is passed on to our children (see Deuteronomy 6:7-9).

One distinct advantage we have over the people that we read about in Deuteronomy is that we have the written Word of God. We have God's words which are eternally settled in heaven and are given to us. The Bible is *the* Authority for living today. However, my own experiences of God's work in my life are not written down. What God did for me last week, last month, or earlier this year on an evangelistic outreach in New York City needs to be passed on to my kids! I suspect that if you will take time today to consider what God has done, you will find plenty to pass on to your kids. Give it a shot today when you are all together.

WHAT'S ON THE MENU?

DAY 39

Deuteronomy 28:1 And it shall come to pass, if thou shalt hearken diligently unto the voice of the LORD thy God, to observe and to do all his commandments which I command thee this day, that the LORD thy God will set thee on high above all nations of the earth:

Have you ever wished that you could wake up each morning and choose the type of day you would have from a menu? "I'll have the 'good day' with an extra side of health and wealth … and super-size it!" Unfortunately, that is just a far-fetched dream! The kind of day you will have is not your choice. There is an old saying, "If wishes were diamonds, beggars would be kings."

Though you cannot choose your circumstances today, you *can* choose to walk with or without God. Even if you walk without God, you are still obeying someone or something. The question you answer daily is, **will I obey the God Who made me?** There are consequences for this choice you make every day.

Deuteronomy 28 is an interesting passage of Scripture. The first fourteen verses are filled with blessings and "good things." The next *fifty* verses are full of judgment and consequences for disobedience. The weight of the whole chapter hinges on the tiny two-letter word *if* in verse 15! One of the consequences for not obeying God was being turned to *"gods* [of] *wood and stone"* (verse 36). Of all the consequences, serving a god made from wood or stone is the worst! How could your day have any significance if you are following that kind of god?

The menu you have to choose from today is not "good day" versus "bad day"; the menu is obeying God or following something/someone else. Any other god you might serve today is probably not made of wood or stone, but the choice before you today is the same choice that was "on the menu" in Deuteronomy

28. Will you choose God or something else? Will you choose God or yourself? The next time you are glancing at a menu, remember that **you can't choose your day, but you *can* choose who you obey.**

◇◇◇

DAY 40
FAITH AND ACTION

Deuteronomy 29:29 The secret things belong unto the LORD our God: but those things which are revealed belong unto us and to our children for ever, that we may do all the words of this law.

If you could call three people, present or past, and ask them each a question, who would you call and what would you ask them? I don't know that this would be my top pick, but I would be curious to ask Amelia Earhart where she landed. I would love to ask Ramesses how in the world they constructed those pyramids. I would even like to talk to Warren Buffett and ask him how he will invest his disposable income this month. Okay, perhaps not the last one!

In any event, my curiosity about another person doesn't come close to the questions I might ask God Almighty if I had the chance. There are plenty of questions that are mysteries to me. Anyone with a curious mind can torment themselves either by incessantly asking questions God has not answered or feigning ignorance to questions that God has already answered. Worrying about things that God hasn't revealed is not good, and giving oneself an alibi for inaction when God has clearly spoken is not good either!

As a child of God, you ought to be living a life of faith and action. You do not need a balance of faith and action; **you need a *harmony* of faith and action.** That is, you need **100% faith**, trusting God all the time, and **100% action**, acting when God commands.

There are some things that God has chosen not to reveal. If He hasn't revealed it, I don't need to know it! I can trust Him about what I don't know. On the other hand, what God has revealed and said I ought to do. I can trust and obey Him. *"The secret things belong unto the LORD our God: but those things which are revealed belong unto us and to our children for ever,* **that we may do** *all the words of this law."*

Don't let what you do not know stop you from obeying what you do know. Trust Him for what you don't know and act on what He has already revealed. God will always give you the light you need in order to do what He wants you to do.

◇◇◇

FOGGY OR FRESH?

DAY 41

Joshua 5:9 And the LORD said unto Joshua, This day have I rolled away the reproach of Egypt from off you. Wherefore the name of the place is called Gilgal unto this day.

As I walked out the door the other day, I mentioned to my wife that the day was going to be a nasty one. It was gloomy outside with a dense fog hovering near the ground. At that point in time, I couldn't look on the bright side because there was no bright side! Ironically, the forecast called for a beautiful Tennessee spring day. Once the sunshine burned through and the fog rolled away, it was exactly the gorgeous day they said it would be!

Perhaps you feel like you are living under a fog, and the forecast for your day is unfavorable. I am so glad that we serve a God Who can roll the "fog" away! In this passage, the children of Israel left Egypt behind and began a new journey into the Promised Land.

A fresh commitment and surrender to God precedes a fresh provision from God. Sometimes we want something new from God for the day, but we are not willing to let go of things from yesterday.

Surrender to God is not a "blank check" for the rest of your life; **surrender to God should be a current commitment that is fresh each day.** Don't let the fog of past successes or failures drift in and cloud your service to God today. God provides naturally and supernaturally, but never forget that God is the One Who provides it all. God frees from the past those who surrender their future. Is the forecast for your day foggy or fresh?

◇◇◇

DAY 42

UNCHANGED

Joshua 6:27 So the LORD was with Joshua; and his fame was noised throughout all the country.

Do you like new things, or are you afraid of them? I guess that depends on whether you are talking about a new car or the first day of school! For Joshua and the children of Israel, everything was new. Moses was dead; Joshua was the leader now. The Promised Land was to be possessed; it was still just a promise. We think of Joshua in all his fame and glory, but that was Joshua after God worked through his life. Joshua, in reality, was just like you and had a mix of excitement and fear about new things.

All of us have a natural fear of new things, even if what is new is good. There is just something intrinsically scary about the unknown! We learn from reading the story of Joshua that **nothing new to us is new to God.** Isn't that wonderful?

The promises that God had made to Moses (Joshua 1:3-9) were new to Joshua, but they were not new to God. God wasn't planning the future—He was already there! The best we can do is only to plan for the future. God is already there waiting for you! Repeatedly, God told Joshua, "I was with Moses ... so will I be with you." **When you take "I was" and add "I will," you get "I AM."**

Do you remember the first day of school? (For some of us, kindergarten was a long time ago!) Your mom thought it was cute; you thought it was terrifying! When you "grow up," your health, work, and future can be just as scary. Remember that God has a track record—nothing new to you is new to God. God can help you; He has helped scores of people before you (like Joshua). The same God that was working in Joshua is the same God that is unchanged today. Nothing in this world ever stays the same; and nothing in God ever changes. In a world of constant change, God is the only One Who never does!

GOT PROBLEMS?

DAY 43

Joshua 9:14 And the men took of their victuals, and asked not counsel at the mouth of the LORD.

You can expect problems in life. Isn't it encouraging this morning to be reminded that you can expect problems? You probably don't need me to tell you that! The story of Gibeon's league with Joshua illustrates the following truths to us:

First, **don't expect your problems to come one at a time.** Joshua's enemies *"gathered themselves together ... with one accord."* (verse 2) Joshua didn't have the luxury of facing each enemy on his own. They came at the same time, and they were united against one man and nation! Wouldn't it be great if your problems in life took numbers, and you could call each one individually when you were ready to deal with it? Problems don't come one at a time. Like the old saying goes, "When it rains, it pours."

Secondly, **don't expect your problems to play fair.** The Bible says that the inhabitants of Gibeon *"did work wilily, and went and made as if they had been ambassadors...."* These next-door neighbors fooled Joshua into thinking they had traveled a great distance. They were seeking a treaty with Israel, but they were being deceptive and were not playing fair. In life, your problems will not play fair either. If life isn't fair, certainly problems and difficulties will not be fair.

What does this all mean for you? **Realize that you need God desperately!** The crux of the story is found in verse 14: *"And the men took of their victuals, and asked not counsel at the mouth of the LORD."* Joshua's real problem was that he didn't ask God for help! Most likely, the big and obvious problems you face are not the ones that will bring you down; it will be the small, subtle, seemingly insignificant problems! You need God to give you strength against the big threats and wisdom to deal with the subtle threats.

Whether your problems are large or small, you need God today. Your problems probably will not come individually and quietly, and they will not play fair. When you face problems in life, recognize your need for God to guide you. And by all means, ask Him for help with the problems!

DAY 44 — ARE YOU UNDER AUTHORITY?

Joshua 11:15 As the LORD commanded Moses his servant, so did Moses command Joshua, and so did Joshua; he left nothing undone of all that the LORD commanded Moses.

The authority you need in life is found in submission to God. A submissive servant has more power than a king acting in rebellion. We are living in a day when our quest for the top has become a race to the bottom. Everyone wants to have a position that gives more authority. The problem is, we too often forget that we are *all* under submission.

The key to possessing authority is not being more assertive, bossing more people, or delegating; the key is submission. **The more authority you are under, the more weight you carry.** You may be a "featherweight," but the more authority to which you are submissive, the more of a "heavyweight" you become! On the other hand, the more you try to make yourself a "heavyweight" and reject others' authority, the more of a "featherweight" you really are.

In Joshua 11, we see a servant, Joshua, acting under orders from Moses, who was God's servant. Joshua's submission to authority provided direction and answers for decisions. **There is only one ultimate authority in the world**, and that is God. He delegates His authority to people, and it is best received with a tender heart and open mind. The more authority you are under, the more weight you carry.

"As the LORD commanded Moses, so did Moses command Joshua, and so did Joshua…."

◇◇◇

TRUE TEAMWORK

DAY 45

Joshua 12:6 Them did Moses the servant of the LORD and the children of Israel smite: and Moses the servant of the LORD gave it for a possession unto the Reubenites, and the Gadites, and the half tribe of Manasseh.

Have you ever made a promise that you had forgotten about until someone came to "cash in" on it? Sometimes promises are easy to make and hard to keep! Two and a half tribes of Israel promised to help everyone else conquer the land before settling themselves on the other side of Jordan (Numbers 32). Their story of conquest in Joshua 12 reminds us that teamwork is important.

Teamwork is not satisfied until individual successes contribute to a common goal. It would've been easy for the tribes of Reuben, Gad, and Manasseh to rest on their laurels and check out when their portion of land was conquered. Instead, what they agreed to do was conquer a land they wouldn't own, build houses they wouldn't live in, and farm lands that they wouldn't enjoy the fruit of. They were part of God's people fulfilling God's command.

Now, we are not Israel, and you are not going to conquer any lands today. But the application is clear, isn't it? Teamwork in any given situation (local church, ministry, job, home, etc.) is not satisfied until your individual success contributes to a common goal. There is nothing wrong with having individual responsibility, but do not let "my" part distract from the goal of the "team."

Teamwork is the context for keeping your word. Moses kept his word concerning the inheritance for these two and a half tribes, and the tribes kept their word to fulfill their responsibility to the other Israelites. Keep your word and remember your responsibilities!

If you have common responsibility, that is communism. If you have only individual goals, that is chaos. **No matter who you are, what you do, or where you serve, never forget that teamwork needs a common goal *and* individual responsibility.**

◇◇◇

DAY 46 — FULLY FOLLOWING

Joshua 14:8 …but I wholly followed the LORD my God.

Earlier this spring, a man named Jeff Smith was removing a stump with his tractor when the tractor unexpectedly flipped, pinning him to the ground. The closest help to him was his teenage daughters, but they were obviously overmatched by the 3,000-pound tractor. Well, they did what they could do and amazingly lifted the tractor off of their dad just enough so that he could wiggle himself free.

Isn't that incredible? A teenage girl and her sister moved a tractor and saved their dad's life! In the Bible, we read about Caleb, a man who was not a teenager—he was an old man! His story is just as amazing.

In Joshua 14, we read of Caleb's effort. Just like Jeff Smith's daughters, **Caleb gave 100% of what he had.** Caleb said, *"I wholly followed the LORD my God."* There is a difference between following and "wholly following." You can seek the kingdom of God, but it is a different matter to *"seek ye first the kingdom of God."* (Matthew 6:33)

We not only read about Caleb's effort, we read about his time. **There is a big difference between giving 100% effort one day**

and giving 100% effort your entire life. Caleb gave 100% effort 100% of the time. He ended his life well.

God continues to give you life for a reason. That means you should continue to live your life for Him. Do you live with purpose? Are you completely following the Lord?

THREE KINDS OF PEOPLE

DAY 47

Joshua 21:45 There failed not aught of any good thing which the LORD had spoken unto the house of Israel; all came to pass.

There are three kinds of people in the world: **promise *makers*, promise *breakers*,** and **promise *takers*.** Which title best describes you? Are you a promise maker? It's not wise to make many promises! The more promises you make, the harder you make it on yourself. Are you a promise taker? A promise taker relies on the promises God has given. Are you a promise breaker, or do you keep your word when you make a promise?

The ultimate promise Maker is God Himself. He never makes a promise that He doesn't remember, have the courage and foresight to follow through on, or have the power to enforce. God is a promise Maker!

Has there ever been a time when you've kept your word, but it didn't appear that way to someone else? Keep in mind that the more you keep your word, the better your track record, and the more slack people will cut you. On the other hand, no matter how great you are, there are always going to be people who dislike you and do not think the best of your intentions. I wish this were not so, but I know it to be true. The best way to respond to folks like that is to keep your promises. Don't be a promise breaker!

To be a promise taker, you must know what God has promised and then act on what He has promised. Joshua 23 is full of promises that God had made to the children of Israel. In the midst of Joshua's challenge to the people of Israel about conquering Canaan, you read phrases like *"as the LORD your God hath promised you"* several times throughout the chapter. God's promises are meant to be acted upon. Faith itself is not what is valuable; Who you place that faith in gives the faith value!

Because God is a promise Maker and always keeps His word, strive to be a promise taker today. Take comfort in the fact that God is already in the future. He owns the future; it is in His hands! Your actions today should be based on what God will do. **Are you a promise maker, a promise breaker, or a promise taker?**

DAY 48

HIS-STORY

Joshua 24:15 And if it seem evil unto you to serve the LORD, choose you this day whom ye will serve; whether the gods which your fathers served that were on the other side of the flood, or the gods of the Amorites, in whose land ye dwell: but as for me and my house, we will serve the LORD.

Who would you say figures most prominently in Israel's history? Joshua? Moses? Abraham? Jacob? Well, by just reading the first thirteen verses of Joshua 24, you cannot help but find the answer! If I counted correctly, you find the personal pronoun *I* seventeen times in those verses; the *I* in those verses is the "I AM." *God* figures most prominently in Israel's history!

Who is the most prominent person in *your* history? **If you can't see His story in your history, you are not likely to follow God in the future.** That is the obvious point of Joshua 24! We are living in a day when, like in Joshua's day, it seems *"evil ... to serve the LORD."* People are calling evil "good" and good "evil." If you believe what the Bible says, you will be labeled as bigoted, hateful, and evil. However, the choice of who to serve is your decision alone—you have to decide. If you fail to see

God in your own history, you will likely fail to follow Him both now and later.

I love to ask people how they came to faith in Jesus Christ. You may not have been saved from a life of ruin and sin, but don't overlook the obvious working of God in your life. When you leave God out of your history, you are swayed to follow something or someone else. Joshua called on the people to decide who—not if—they would follow. **If you are going to follow God in the future, you must see His working in your past.** That is what Joshua did in chapter 24, and that is what we must do as well.

Consider writing down or telling others what God has done in your life. If I were you, I would start with the people in my house and under my care! Is His story a part of your history?

◇◇◇

ALMOST

DAY 49

Judges 1:28 And it came to pass, when Israel was strong, that they put the Canaanites to tribute, and did not utterly drive them out.

Almost is probably the biggest word in the Bible, if you think about it. Is there a difference between winning and *almost* winning? There is a huge difference! In the book of Judges, *almost* is the difference between the blessings of complete obedience to God and the turmoil of not utterly driving out the Canaanites. In the matter of salvation, *almost* is the difference between heaven and hell. King Agrippa said to Paul, *"Almost thou persuadest me to be a Christian."* (Acts 26:28) **The word *almost* has great consequences, but it holds no weight in God's court!**

Are you winning, or are you *almost* winning? Are you obeying, or are you *almost* obeying? The children of Israel chose compromising instead of conquering. Although this compromise was more lucrative, ***almost* obeying was simply delayed defeat.** Ultimately, Israel was defeated by women, not warriors; by religion, not heathenism. In other words, Israel set themselves up for a prolonged defeat. They were investing in something that would ultimately conquer them!

In stark contrast to *almost* obeying, we find that Caleb *"expelled thence the three sons of Anak."* (verse 20) The sons of Anak were not wimpy weaklings—they were giants! (And Caleb wasn't exactly a spring chicken, either!) Caleb's complete obedience was not because he had more ability than the other Israelites; the difference was that he *"wholly followed the LORD."* (Joshua 14:8) **There is a profound difference between *almost* following God and *wholly* following God!**

The key to completely obeying versus almost obeying lies in the decision you face today to obey or to compromise. In what area of your life are you most likely to *almost* obey? That is the area that you are *almost* going to win! May we be the "wholly" kind of Christians, not the "almost" kind of Christians!

◇◇◇

DAY 50 — NO GOOD ANSWER

Judges 2:2 And ye shall make no league with the inhabitants of this land; ye shall throw down their altars: but ye have not obeyed my voice: why have ye done this?

As kids, my sisters and I sometimes committed crimes that came with mandatory sentences! They were called "spanking matters." If you violated one of the "spanking matters," there was no doubt about what was going to happen. You could plead innocence, insanity, or blame your sisters (all of which I tried); but the sentence was unavoidable. Dad would always ask a series of very cruel questions which informed me of what was about to happen. The conversation went something like this:

"Wil, did you do it?"

"Yes, sir."

"Did I ask you not to do it?"

"Yes, sir."

"What did I say I would do if you did it?"

You know, there is no good answer to that last question! My dad used that question to inform me, not himself. The Lord used such a question in Judges 2 to inform His people. He asked, *"Why have ye done this?"* There was no good answer because there was no good reason!

Oftentimes we excuse sin by reasoning, "I'm being overlooked," "They did me wrong," or "I deserve it." Do you think those excuses hold weight with a holy God? If you honestly answer the question, "Why have I done this?", you will often find that your answer holds no weight before God. **There is never a good answer to the question of why we sin!**

Which is a better excuse—sinning because you weren't thinking ("it just came naturally") or sinning because you *did* think and chose to do it anyway? Neither excuse is a good answer! The question God asked Israel in Judges 2 is worth our pondering this morning. *"Why have ye done this?"* The question "Why?" is best pondered before—not after—you make any decision today.

◇◇◇

ARE YOU A HERO?

DAY 51

Judges 6:12 And the angel of the LORD appeared unto him, and said unto him, The LORD is with thee, thou mighty man of valour.

Every day must have felt like a Monday to Gideon. Can you imagine if you were in Gideon's sandals? He was overwhelmed by his circumstances. He was overshadowed by his own family. He was underpowered when compared to the Midianites.

I think Gideon, like many people, was waiting for a hero to swoop in and answer all his concerns. **Sometimes God doesn't send a hero; He makes a hero.** Instead of waiting for a hero, maybe *you* are the hero! God will enable you to do what only He can do.

God receives the greatest glory when He does the greatest things through the smallest people. That was the theme of Gideon's life! The children of Israel needed a hero, but the hero was already there. Gideon was not mighty, not much of a man, and not valiant, yet God said, *"The LORD is with thee, thou mighty man of valour."*

You will never know what your potential is until you have problems. Gideon didn't know that he had everything for which he was praying. Valor? He had it because God was with him. Might? He had it because God had sent him.

When you ask God for a hero, realize that God may be sending you. Be willing to be the person you wish God to send. Be willing to take the action you wish someone would take. You may be the hero you are looking for!

◇◇◇

DAY 52 — WAGING WAR ON DIFFICULTIES

Judges 7:2 And the LORD said unto Gideon, The people that are with thee are too many for me to give the Midianites into their hands, lest Israel vaunt themselves against me, saying, Mine own hand hath saved me.

Have you ever been up against something that was bigger than you are? Have you ever faced opposition that you could not overcome by your own strength? If so, I have good news! You are not the first person to face such odds, as the story of Gideon in Judges 7 reminds us.

The Art of War, by Sun Tzu, is one of the oldest books still in print. Although a book about war, its principles have been adopted by business people around the world. *The Art of War* is a book of strategy, and it explains how to use terrain, trickery, and your enemy's strengths to your advantage. Long before Sun Tzu wrote the book, Gideon actually lived it out!

Gideon reminds us that in waging war on your difficulties, **it is crucial to recognize the source of your strength and to employ that strength correctly.** He literally had too much strength (men) for God to use him. In the end, Gideon and his army of three hundred didn't conquer the Midianites; God did. The question is not, "How much strength do you have?" The question is, "Where is your strength coming from?" Never forget that God is the Source of your strength!

Sometimes when you are at your strongest, you are at your weakest. It is not enough to just have strength; you must employ that strength correctly. Take the game of golf for example. If you are like me, the harder I try to hit the ball, the worse my game is. On the other hand, I know a twelve-year-old boy who can outplay and outswing his dad and grandpa. A golfer with only "boy strength" can outplay a golfer with "man strength" when he employs his strength correctly.

God gave Gideon the victory with only three hundred men, some pitchers, and some trumpets. He literally used Israel's weakness to employ His power, and He turned the Midianites' strength against them. As you face today, Who is the Source of your strength? How are you employing your strength?

◇◇◇

PROBLEMS WHILE PURSUING

DAY 53

Judges 8:4 And Gideon came to Jordan, and passed over, he, and the three hundred men that were with him, faint, yet pursuing them.

When I read Judges 8, I cannot help but remember a message preached in college chapel one spring. We were weary college students and needed encouragement about being faint yet still pursuing the objective God had given us. I really needed that!

It is interesting to take note of all the groups of people that Gideon encountered as he pursued the Midianites. The truth is, you learn about other people by the way they act toward you, but **you learn about yourself by the way you react to them**. Let's look at a few groups of people Gideon met along the way:

1. The men of Ephraim (verse 1) were mad because Gideon didn't call them when he went to battle. The battle was over and the victory was secured, but these guys had **wounded pride.**

2. The men of Succoth (verse 5) didn't want to give bread to Gideon's army before the battle was over. They were **uncommitted people**. Gideon needed help, but the men of Succoth were not willing to risk something of their own.

3. Jether, Gideon's son (verse 20), was a **reluctant partner**. Gideon was depending on his son, but Jether *"drew not his sword: for he feared, because he was yet a youth."* It is tough to continue pursuing when a partner won't "pull the trigger"!

4. The men of Israel (verse 22) had a **flattering perspective**. But being king was not what Gideon was pursuing. Have you ever been flattered? Don't believe it! If you are easily flattered, you will be easily discouraged.

Yes, Gideon was faint. Yes, Gideon still pursued. Proverbs 24:10 says, *"If thou faint in the day of adversity, thy strength is small."* What does your response to people while "pursuing" say about you? Don't let problems distract you from the objective God has given you!

ENDING WITH HONOR

DAY 54

Judges 9:54 And he called hastily unto the young man his armourbearer, and said unto him, Draw thy sword, and slay me, that men say not of me, A woman slew him....

This spring, a man jumped from the Empire State Building and only suffered a broken ankle. How was this possible? Well, he leapt from the top floor but landed on the next floor down. Talk about adding insult to injury!

This is not a pleasant topic to discuss, but how would you choose to die? Your answer is probably, "Wil, I don't want to die!" That is fine, and I would agree. Most of us do not have a specific way we want to die, but all of us want to end honorably.

Ending honorably is not determined by the way you die; it is determined by the way you live. You don't choose to die; you will die (Hebrews 9:27). You can, however, choose how you will live.

Abimelech was a man consumed with his own honor. He did whatever it took to get want he wanted from people. Ironically, a man possessed with building himself and destroying everyone else was done in by a woman and a rock! He "fell on the sword" (with an assist from his armourbearer) to keep from being killed by a woman.

Abimelech didn't die dishonorably because a woman dropped a rock on his head; he died dishonorably because he *lived* dishonorably. The truth is, **you cannot end honorably if you don't live honorably**! Most people try all kinds of antics to hold on to their sense of self-importance. Don't seek to be promoted; seek to serve the best you can where you are!

Seeking to be prominent is shallow and pointless. If you seek to be productive, God can take care of promoting you to the most productive place for you. Remember, *ending* with honor requires *living* with honor today.

DAY 55

STRENGTH OR WEAKNESS?

Judges 14:7 And he went down, and talked with the woman; and she pleased Samson well.

This morning, we are going to look at the life of the weakest man in the Bible, Samson. "*Weakest*?! You mean *strongest*, right, Brother Wil?" No, I really did mean to say *weakest*. The source of Samson's strength was not his hair or his wit; his strength came from *"the Spirit of the LORD."* However, all the references to Samson's life in Judges 14 were not about God's Spirit; they were all about Samson's flesh. It is ironic that the strongest man in the Bible is largely defined by his weakness! One could argue that *weakness* better defines Samson than *strength*.

Samson's life teaches us that **relying on your own strength invites the devil to play to your weakness.** We all have strengths and weaknesses. No matter how strong you may be, you also have a weakness. In fact, oftentimes the stronger your strengths, the stronger your weaknesses. History is full of examples of people who were powerful financially, politically, or intellectually but had mammoth weaknesses that defined their lives.

The devil would love nothing more than to play to your weaknesses so that you are not able to be defined by your strengths. What can you do to keep from being defined by your weaknesses? First, **remember that you are never too old for submission to authority.** Samson's great weakness was actually submission to authority—not women, wealth, or wit. He was not submitted to his parents when it came to the woman of Timnath, and that weakness was loaded with heartache and trouble. You will never be stronger than your need for submission.

Secondly, **remember that you are never stronger than the Holy Spirit** who gives you strength. Samson was witty, strong, and sociable; but none of those strengths could replace the need for God's strength by God's Spirit. In the end, Samson lost his hair, his strength, and his freedom; yet he tragically *"wist not that the LORD was departed from him."* (Judges 16:20)

Your weaknesses can be minimized when you realize the importance of authority and dependence upon God for strength. When you rely on *your* strength, the Devil will play to your weakness. That truth is the reason Samson is largely defined by his weakness instead of his strength. Will your life be defined by your strength in God or by your weakness?

IN-LAWS

DAY 56

Ruth 1:16-17 And Ruth said, Intreat me not to leave thee, or to return from following after thee: for whither thou goest, I will go; and where thou lodgest, I will lodge: thy people shall be my people, and thy God my God: Where thou diest, will I die, and there will I be buried: the LORD do so to me, and more also, if ought but death part thee and me.

I have a great story about a mother-in-law. (Usually, that statement does not precede something flattering!) When you read the book of Ruth, you find a very sweet story about a woman named Ruth and her mother-in-law Naomi. Nothing seemed to be good in Naomi's life: no blessing from God, no food, and no husband. What could possibly be sweet in all of this?

Enter her daughter-in-law, Ruth. Ruth's husband (Naomi's son) died, and now Naomi releases Ruth to go back to her people and her home, just as she would do. Ruth's promise in verses 16-17 is notable and well-known, often quoted in marriage ceremonies. I think her promise is a wonderful sentiment for a bride to say to her new husband, but this promise goes further than just a wedding ceremony!

This promise is ironic. While brides often quote this passage to their grooms, it was originally made to a mother-in-law! Imagine a bride saying this to the groom's mother. That kills the romance, doesn't it? Now imagine saying this to *your* in-laws: "Where you go, I will go; where you live, I will live; …."

This promise is profound. Why? Nothing could be more true than saying to your in-laws, "Thy people shall be my people." I think about my own home. The Wil Rice home today is really a combination of my wife's family and background and my family and background. Our home and our kids are a blend of two distinct families. I am responsible to lead and guide my home; but if you ever visit, you'll see that our food—even our soap—is distinctly Birky! My wife brought all of those things to our home when we were married.

This promise is instructive. Though we may not actually repeat this promise to our in-laws, it would do us good to imagine saying it to them. You are not doomed to your past or your spouse's past, but your in-laws can inform you of what to expect for years to come. Furthermore, whether you like it or not, *marriage* spells "duty." We have a God-given obligation to honor our parents; when you get married, you multiply that obligation!

When thinking about families and in-laws, I am thankful for God's grace. Neither Ruth nor Boaz had a great heritage (Ruth was from Moab; Boaz descended from Rahab), but God clearly turned something bad into something beautiful. **The story of Ruth is ultimately a story of God's grace**, but the story also has great instructions for us when it comes to marriage and our in-laws!

DAY 57

WHO WILL GOD USE?

Ruth 4:17 And the women her neighbours gave it a name, saying, There is a son born to Naomi; and they called his name Obed: he is the father of Jesse, the father of David.

What kind of person will God use? A person who is perfect? A person who is smart? A person who has great ability? That is an important question, and the answer is illustrated for us in the book of Ruth. Ultimately, the story of the book of Ruth is not about Ruth, Naomi, or Boaz. At the end of the book, you find the genealogies of the Lord Jesus Christ!

The royal line of David, from which the Messiah came, is intriguing. You find a bunch of flawed people! Pharez was a man born from a shameful relationship (Genesis 38). Boaz was a descendant of a harlot (Rahab). Ruth was a Moabitess. These were all unlikely players in the genealogy of King David and the Messiah who came!

God uses faithful people who are flawed, but He will not use flawed people who are not faithful. In Ruth, all the characters had pasts that were flawed, but they were faithful in the time that God granted to them. There is not an excuse for wickedness, but God will still use a faithful person.

Each of us has a flawed past. If you look hard enough, you can probably find things in your life and the lives of your ancestors that are neither honoring to God nor encouraging to you. However, if you are going to serve God, you must serve Him in the time He has given *you* and not dwell on your flawed past. Being flawed is our nature; being faithful is by God's grace. In the daily decisions you make today, don't let your flawed past keep you from being faithful to God!

◇◇◇

ENDING AND BEGINNING

DAY 58

Ruth 4:22 And Obed begat Jesse, and Jesse begat David."

Have you ever skipped to the end of the book you are reading? You might do so to find out who is revealed or how things work together. Some books end with, "The end," or that famous fairy-tale quote, "And they lived happily ever after." Well, the book of Ruth is a great book to skip to the end because **the ending is really the beginning**!

The very last verse in Ruth 4 says, *"And Obed begat Jesse, and Jesse begat David."* Why is this important? Because the lineage of the Lord Jesus Christ goes through David. Each person in those long lists of genealogies that we like to skip is a real person with real heartaches, triumphs, joys, and difficulties.

The beginning of Ruth's story is actually the end of Naomi's story. Naomi's husband died, her two sons died, and she was left in a strange land surrounded by strangers. Little did she know that God had a plan for her and her daughter-in-law, Ruth. **Sometimes the end of our plans marks the beginning of God's.**

God may still be writing a chapter in your story at the point when you think you have finished the last chapter. God is never done with you until He takes you to heaven. Until then, don't stop!

When Elimelech died, God had the world in mind. When Naomi was coming to the end of her plan, God was coming to the beginning of His. **God always has a perfect perspective, and you can (and ought to) trust Him.** Sometimes the end of your story is actually the beginning of the story that God intends!

DAY 59 — PLEASING OR APPEASING?

I Samuel 4:3 And when the people were come into the camp, the elders of Israel said, Wherefore hath the LORD smitten us today before the Philistines? Let us fetch the ark of the covenant of the LORD out of Shiloh unto us, that, when it cometh among us, it may save us out of the hand of our enemies.

I Samuel 4-5 tells the story of Israel's battles with the Philistines during the time of Samuel. At the first battle, Israel's army was soundly defeated. Instead of looking at their own lives for the reason for the loss, they looked for a good luck charm! The elders said, *"Let us fetch the ark of the covenant of the LORD out of Shiloh … that … it may save us out of the hand of our enemies."* They were more concerned about an object (the ark of God) than the God of the object!

Of course, with the ark's return, Israel was expecting victory as the Philistines braced for defeat. Ironically, the exact opposite happened! Israel was defeated again—and the Philistines even took the ark back with them. The pagans figured Israel's good luck charm could only help their wicked causes thereafter. The Israelites were more distraught about losing their "good luck charm" (the ark) than they were about living without the blessing and favor of God!

Truthfully, Israel treated God just like the Philistines did—like a "good luck charm." There is a danger in limiting God to places, human methods, or historical successes. **When you put God "in a box," you limit a God that is unlimited!** Sometimes, we show God the very same lack of reverence that we find in I Samuel. Often we are concerned that God is on *our* side; we should be concerned that we are on *God's* side.

There is a fundamental difference between *pleasing* God and *appeasing* God. God is not some "good luck charm" you can manipulate. No sir, you do not appease God in order to control Him; you please God by letting Him control *you*. In other words, when you were saved, you received all of the Holy Spirit that you will ever get. The question is, does He have all of you?

God Almighty does not want to be appeased by what you can give Him; He wants to be pleased by what He does through you. Both Israel and the pagan Philistines treated God like a "good luck charm." Is your life the same way, seeking to appease God? **Are you pleasing Him by letting Him have complete control?**

◇◇◇

DO WHAT YOU CAN

DAY 60

I Samuel 14:6 And Jonathan said to the young man that bare his armour, Come, and let us go over unto the garrison of these uncircumcised: it may be that the LORD will work for us: for there is no restraint to the LORD to save by many or by few.

Most big things—things worth doing—cannot be done at one time or by one person. This is often true, but there are exceptions to this truth. Instead of allowing this truth to be an excuse for discouragement and inaction, the exceptions should be motivation for *taking* action even when you find yourself alone. In I Samuel 14, Saul was basically sitting on his hands. He was sitting under a pomegranate tree in Gibeah, waiting for something to happen. After Saul and his army had provoked the Philistines, all the brave men of his army headed for the hills and the holes!

Jonathan, Saul's son, was a brave man, a man of action, faith, and selflessness. His story is an example of someone who just does what he can. Sometimes we think that if we can't do everything, then we shouldn't do anything. That's not the case! **You cannot do everything, but you can do something.** You can't do everything, but God can! If you will do what you can do, God will complete what is beyond your grasp.

Jonathan and his armor bearer took action, and God enabled them to defeat the Philistines while Saul and others sat on the sidelines. How did God do it? First, God gave Jonathan and his armor bearer grace and strength ten times their number when they took action. Secondly, God sent an earthquake (verse 15). When Jonathan did what he could, God did what only He could do. Thirdly, God used the Israelites who were among the Philistines to turn against the Philistines and fight. The Philistines were getting hit from within, without, and all around!

The end of the story is found in verse 23: *"So the LORD saved Israel that day: and the battle passed over unto Bethaven."* God saved Israel when a man (Jonathan) took action, a deed which eventually led others to join in the fight. Jonathan had the courage to follow God and lead others. Today, don't be discouraged and sit idly by when there is action to take. Follow Jonathan's example and do what you can, when you can.

USING PEOPLE

DAY 61

I Samuel 15:15 And Saul said, They have brought them from the Amalekites: for the people spared the best of the sheep and of the oxen, to sacrifice unto the LORD thy God; and the rest we have utterly destroyed.

There are some people that never think about anyone else but themselves. The only thing worse than that type of person is the person who is always thinking of other people but only in terms of what he can get from them! I suppose that none of us want to use people, but **sometimes we are thinking about ourselves when we think about other people.** How do you know if your dealings with others are all about you?

Well, King Saul was a self-absorbed man that thought of others only in terms of how they could benefit him. In I Samuel 15, God told him to destroy the Amalekites and all that they had. Though the command was clear, Saul did not completely or honestly obey. When confronted by Samuel, Saul piously claimed, *"I have performed the commandment of the LORD."* (verse 13)

Saul had not obeyed the Lord, and even the sheep and oxen admitted as much! He tried to shift the blame to the people. *"They have brought ... **the people** spared the best ... the rest **we** have utterly destroyed."* One way Saul used people was to **buffer his responsibility**. You know you are using people when you try to put them between you and your responsibility.

Saul also used people to **enhance his reputation**. He said, *"I have sinned: yet honour me now, I pray thee, before the elders of my people...."* (verse 30) Sometimes our dealings with other people are not so much about them but about how they can help us or enhance our standing.

If we are honest, we can be just as self-absorbed as Saul was. Even when we try our best to think of others, all we think of is how they can help *us*! Saul is a good example of what *not* to do with other people. Do you use people to blunt your responsibility? Do you use people to enhance your reputation?

DAY 62 — HOW DOES YOUR WORLD LOOK?

I Samuel 21:10 And David arose, and fled that day for fear of Saul, and went to Achish the king of Gath.

Think about the people in your life. Out of all those people, only a handful are probably invested and interested in you. Do you like the people in your world? The better question is, do you like the kind of people you are attracting to your world?

King Saul looked around and saw people who were dishonest, self-serving, and cut-throat. They were exactly like him! He had driven off the people who were loyal, brave, and selfless. We find that David fled to Gath—the enemy—to escape Saul. David began acting like an enemy because that's the way Saul treated him!

How are the people in your world? We tend to **get what we give**. We tend to **get what we expect**. We tend to **attract what we reward** and **repel what we punish**.

As a parent, you will not get what you want; you will get what you are. You will not get what you want; you will get what you give. You will not get what you want; you will get what you reward. That is true in a marriage, in a home, and in a country.

What are you giving? What are you expecting? What are you rewarding? If you don't like the way your world looks, the reason may be *you*!

IT'S NOT FAIR!

DAY 63

I Samuel 30:23 Then said David, Ye shall not do so, my brethren, with that which the LORD hath given us, who hath preserved us, and delivered the company that came against us into our hand.

Have you ever heard someone say, "That's not fair"? More importantly, have *you* ever said, "That's not fair"? Whether it's a kid or his parent saying (or thinking) it, what's fair is never fair. Someone always has it easier, better, or simpler! The grass is always greener on the other side!

Cries of "That's not fair!" are exactly what greeted David after the trouble at Ziklag and the subsequent battle. Two hundred of David's men were so weary that they had to "stay by the stuff" near a brook. When the army recovered all of the plundered goods along with the spoils of the battle, the four hundred men that fought on the front line didn't want to share with the two hundred men on the sideline. Why should they share when the weak men didn't even help? Can't you almost hear them say, "But that's not fair!" David's solution was to divide the spoils, graciously sharing with the two hundred weary soldiers.

When you are tempted to think or say, "That's not fair!", remember that life is not fair. Don't lose sleep over something that is too much or too little, too easy or too hard, too in front or too behind the scenes. Remember, first of all, that **what you have is given by God.** Nothing you have is really "yours"; God owns it all and gives you whatever you have, whether that is little or much.

Secondly, remember that **what you have is improved by others.** Whether you are on the front lines of battle, or you are staying by the stuff, what you have can be enhanced by others. If everyone just does what they can, the need is always met. A carpenter can build many things that I never could. On the other hand, a carpenter may never preach at a church or at the Bill Rice Ranch. The point is not making sure everything is "fair"; the point is doing your part with what you've been given. **A common purpose or goal is only met when people take individual responsibility.**

Don't waste your life complaining about what is not fair. Can I let you in on a secret about what you have? It's *not* fair! What you have is given by God and improved by others. Use what you have to do the job God has for you to do.

DAY 64 — TWO STRENGTHS

II Samuel 2:1 And it came to pass after this, that David enquired of the LORD, saying, Shall I go up into any of the cities of Judah? And the LORD said unto him, Go up. And David said, Whither shall I go up? And he said, Unto Hebron.

Who is the strongest person you know? Some people are strong emotionally. They are practically "bomb-proof" when it comes to emotional events. Some people are strong physically. You can't miss their muscles and brute strength! Some people are strong intellectually. Go ahead and try to stump them if you wish, but you will quickly decide that they are leaps and bounds above you intellectually.

In life, there are two types of strength: **the strength of power** and **the strength of knowing how to control power.** There is nothing wrong with having the strength of power, whether that is found in a national army or in the determination and confidence to do what God made you to do. We can all use a little bit of determination and confidence in doing right today! In fact, most of the help you ask God for is "strength of power." There is nothing wrong with asking for that!

However, if the only strength you have is the strength of power, you are in trouble. For example, the most dangerous type of horse is a three-year-old. He is just big enough to hurt you and young enough to be ignorant. A horse with power is good; a horse with power under control is usable. **Power that is not constructive is destructive!**

We need the strength of knowing how to control the force we have. When David faced trouble and confusion at the beginning of his reign, he *"inquired of the LORD."* Knowing how to use the power you have is mightily important! Each person has some power, regardless of how you feel about it. Whatever power you have, **you'll have more if you know how to use what you have**!

You may pray for the right words to say in a given situation. Knowing *how to use* those words makes all the difference. The right words at the wrong time or in the wrong way can be damaging. David knew the importance of power coupled with the knowledge of how to use it. Strength of power is not bad; knowing how to use that power can make all the difference!

◇◇◇

THE MOST HARMFUL SIN

DAY 65

II Samuel 12:7a And Nathan said to David, Thou art the man.

Quite possibly, the most annoying thing in the world is a phone that rings during a public event. Of course, the most embarrassing thing is finding out that it is *your* phone ringing! Perhaps the most annoying noise is a car that squeals and squeaks while driving down the road. The most embarrassing thing is discovering that the noise is not following you; the noise *is* you! Perhaps kids running around like hooligans is annoying. On the other hand, realizing that the hooligans belong to you is embarrassing!

The most annoying thing in the world is sin *in theory*; the most embarrassing thing in the world is sin *in you*. It can be the most helpful to realize, like David, that *"thou art the man."* God cared enough about David to send someone (Nathan) to help him, and God cares about you!

David was completely offended and outraged at the sin in Nathan's story. Like a ringing phone, a squealing fan belt, or rowdy kids, David didn't realize that *he* was the offensive and outrageous sinner! He was content for "the man" to pay for his wrong (verse 5) until he realized that he was "the man" (verse 7).

We are offended by sin in theory and in others, but **sin draws us in by appealing to our need.** Need recognition, possessions, or wealth? The devil can supply all of them, but they will not be right. **The sin that does the most harm is the one you do not recognize.**

Pride in others offends us, but our own pride deceives us. Gossip offends us, and we love to gossip about the person who gossips. Selfishness in others stands out to us, but does our own selfishness stand out? The most damaging sin today is not a theoretical sin; it is the sin that you do not acknowledge.

◇◇◇

DAY 66 — TRUE FRIENDS

II Samuel 15:37 So Hushai David's friend came into the city, and Absalom came into Jerusalem.

Who is your best friend? Your best friend is probably the one who is there at your worst times. Such was the case in II Samuel 15 as David fled for his life. Absalom, his son, was a rebel and had overthrown David's kingdom. The Bible mentions *"Hushai David's friend."* Never was that title more important than this time in David's life!

What defines a friend depends on where you are in life. Until now, King David had many "friends." Several things made these many friends for him: palaces, possessions, power, popularity, and politics. When David went from being a shepherd to being the king of Israel, I imagine the "friends" came out of the woodwork! Everyone wants to be a friend of the king. The bandwagon of popularity was probably full after David killed Goliath. The women of Israel even made songs about the might of David!

While palaces, possessions, power, popularity, and politics all make you friends, *problems* **will reveal your true friends.**

You will not choose problems over popularity or power, unless you realize that problems reveal your true friends. The "friends" deserted David when Absalom overtook the throne. The bandwagon was deserted. No one was singing in the streets while David fled for his life. True friends were worth more than anything David had!

The same principle is true in your life. While you may seemingly have "friends" when life is going well and your popularity and power are at their peak, problems will reveal the true friends who are left standing. **Don't despair over problems**—one thing they will do is reveal who are the true friends! Even when it seems like *no one* is a true friend, **you have a "friend that sticketh closer than a brother" in the Lord Jesus!**

Decide to *be* **a friend when it is needed most.** Hushai was a friend to David. Who are your true friends? Who needs *you* to be a true friend?

◇◇◇

REST ON THE RUN

DAY 67

II Samuel 16:12-13 It may be that the LORD will look on mine affliction, and that the LORD will requite me good for his cursing this day. And as David and his men went by the way, Shimei went along on the hill's side over against him, and cursed as he went, and threw stones at him, and cast dust.

This passage in II Samuel recounts the story of David fleeing for his life. He was on the run from his own flesh and blood, Absalom. As bad as the situation was, with David fleeing for safety, this man Shimei was up "in the grandstands" cursing and casting dirt. David was on the run but needed rest.

Do not forget that David was a warrior—he could have squashed this little "pest" without much trouble. On top of that, one of David's servants, Abishai, was more than willing to take care of the problem: *"Why should this dead dog curse my lord the king? let me go over, I pray thee, and take off his head."* (verse 9)

But instead of taking on people problems, David entrusted other people to God.

David acknowledged his part is verse 10 when he said, *"So let him curse, because the LORD hath said unto him, Curse David. Who shall then say, Wherefore hast thou done so?"* David acknowledged his part, but he also entrusted people to God. In verse 12 he said, *"It may be that the LORD will look on mine affliction, and that the LORD will requite me good for his cursing this day."* David gave his people problem to God, and he trusted that God would requite, or repay.

So what is the key to having rest on the run? David shows us that you must entrust other people to God, believing that He will repay. Today's troubles are big enough without your worrying. But in the midst of troubles, you can find rest the same way David did. Our natural reaction is to return evil for evil, yet David—every bit as human as we are—found rest on the run.

Do you have people problems today? Ask God for wisdom to know what you ought to do. It really is a matter of trusting God—*"the LORD will requite me good…."* Keep in mind that God is the One in control. Psalm 90 may be a great help to you. That psalm reminds us that God takes vengence, is our Judge, and is our defense. **You can trust His timing and His way for the people problems you face today.**

MY NAME IS _____

DAY 68

II Samuel 18:18 Now Absalom in his lifetime had taken and reared up for himself a pillar … and he called the pillar after his own name: and it is called unto this day, Absalom's place.

Do you like your name? It's good if you do! When I was a kid, I had an alter ego with a different name. I liked both of my names! The truth is, **you have two names: the one your mother gave you and the one you will give yourself.**

Everyone wants to be remembered. For what do you want to be remembered? Absalom is an example of someone who had a good name (as David's son) but turned out bad. The name *Absalom* means "father of peace," yet we remember him for being at war with his father. He was very intent on building a name and a monument for himself.

Begin today by deciding what you want your name to mean, and then make every decision in light of that name. People will usually remember you for one impression. What do you remember about Abraham? About Jezebel? About Lot? About Stephen? What is the one thing people will remember about you?

If you have a good name, keep at it for the long haul. If your name is less than desirable, start today by making right choices. **You alone will determine the meaning of your name.** Decisions build character. Character builds a name. A name builds a legacy. What is your name?

DAY 69

A WARRIOR'S GENTLE SONG

II Samuel 22:2-3 And he said, The LORD is my rock, and my fortress, and my deliverer; The God of my rock; in him will I trust: he is my shield, and the horn of my salvation, my high tower, and my refuge, my saviour; thou savest me from violence.

Some people can surprise us by being or doing something that is unexpected. Take, for instance, a professional football player … that takes ballet. He may see it as beneficial and helpful, but you would not expect it! Or perhaps on Sunday morning you see a guy in your church choir, and you think to yourself, "He sings?" In this chapter of II Samuel, we find a surprising song. David—the "bloody man," the warrior—writes a song of dependence on God's help. David was tough, yet tender; and he definitely made the Lord his "rock."

We see David's dependence in verses 2-3 when he uses words like "rock," "fortress," and "horn." A fortress is not an offensive weapon; it is a place for protection, a place you can trust. A horn is a weapon for defense, and David said **God** was his defense.

Not only do we see what God was to David, but we can see what David did in verses 4 and 7. He says, *"I will call on the LORD, who is worthy to be praised: so shall I be saved from mine enemies… In my distress I called upon the LORD, and cried to my God…."* David was a warrior, but he was a warrior who trusted God and prayed for the help he needed. **Never forget that we are always in trouble; sometimes we are just not smart enough to realize it!**

Finally we see *what God does* in verses 8-10. When you read these three verses, you find that God is both tough and tender— He heard David's cries for help, but also *"the earth shook and trembled … because he was wroth…. and fire out of his mouth devoured…."* There is no doubt that God is powerful!

So I think this psalm of David begs an application. You can know that God does pay back. David says in verse 21, *"The LORD rewarded me according to my righteousness,"* and in verse 25,

"Therefore the LORD hath recompensed me." God pays back—what side of the "payback" are you on? That can encourage you today that, like David, the Lord will take care of your trouble. But that can also be troubling if you are on the wrong side of things, and God "takes care" of you!

God lives, avenges, and makes things right. God will reward—what side are you on? I hope that you are on the right side, and if you are, **be encouraged today that you can trust God with your enemies.**

◇◇◇

THREE CHOICES

DAY 70

II Samuel 24:14 And David said unto Gad, I am in a great strait: let us fall now into the hand of the LORD; for his mercies are great: and let me not fall into the hand of man.

Have you ever compared calamities with someone? Maybe you tell someone what happened to you, and they say, "You think that is bad? Listen to what happened to me!" Comparing injuries, bad days, or troubles is a silly thing to do. In II Samuel 24, God allows David to choose one of three judgements for his sin. David had a tough decision to weigh!

What led David to this choice? We find his sin in the first part of the chapter. David numbered the people (verse 1), but he was trusting in himself instead of God. David confessed to the Lord in verse 10, *"I have sinned greatly … take away the iniquity of thy servant; for I have done very foolishly."*

Because of his sin, David was now ***"in a great strait."*** David's choice was to plead for mercy and leave the choice to God. God's mercies *"are great,"* and David could trust God's choice. You, too, can trust God in the way He handles those who wrong you. God will pay back. **We can trust God with our enemies,** and we can trust God to do right His dealings with us when we do wrong.

God knows what is right. He is a just God and a merciful God. If you were in David's shoes, which judgement would you want? The most severe judgement or the one with a little mercy? I would want God's mercy!

After David's sin and his strait, we see his sacrifice in verse 24. He bought the threshing floor and oxen at full price because, as he said, *"I will surely buy it of thee at a price: neither will I offer burnt offerings unto the LORD my God of that which doth cost me nothing."* Interestingly, David's son, Solomon, would later build the Temple on this very place.

David trusted God to do the right thing. Today, you can trust God to do right with your enemies, and you can trust God to do what is right with you. **God's mercies are great, and He will give the right thing in the right measure.**

◇◇◇

DAY 71
GREATER WORK

I Kings 5:5 And, behold, I purpose to build an house unto the name of the LORD my God, as the LORD spake unto David my father, saying, Thy son, whom I will set upon thy throne in thy room, he shall build an house unto my name.

Sometimes we can be impressed by the physical size of a structure. Sometimes we can be impressed by the wealth of a structure's design. In the case of Solomon's Temple in I Kings 5, *both* were true! It was a magnificent temple in every sense of the word.

So let me ask you: Who built Solomon's Temple? Did slaves build it? Did King David build it? Or did Solomon build it? Well, no matter the skill or muscle, no one could have ever built the Temple alone. In fact, Solomon did not have all of the materials to build the Temple! Fine, then. Whose idea was it to build the Temple? That wasn't Solomon, either!

Let's review. Solomon's Temple was David's vision, completed by scores of people with materials from others, including the King of Tyre. The truth is, **any work of God worth doing is bigger than you but should include you.** You must keep a proper perspective. Solomon's work in building the Temple was just as important as the stone workers' work and the tree cutters' work. The Temple was bigger than Solomon, but it would have not been built without Solomon.

Don't have "tunnel vision" in your part of God's work, so that you forget that His work is greater than you are. Likewise, don't neglect your part of God's work! Anything worth doing for God is bigger than you, but—thank God—it includes you!

◇◇◇

LIVING BEYOND YOUR LIFETIME

DAY 72

I Kings 7:51 So was ended all the work that king Solomon made for the house of the LORD. And Solomon brought in the things which David his father had dedicated; even the silver, and the gold, and the vessels, did he put among the treasures of the house of the LORD.

How much of you will they bury when you die? I suppose they'll bury all of you! Now think about how much you invest in your body every day. Soap, deodorant, hairspray, and clothing are not bad things; but they will all be gone when you are. **The only thing that will remain is what you invest in people.**

David wanted to build God a temple. When he could not, he invested in his son, Solomon, who would build it. David left this life, but David left a son. Solomon got all of the treasure for the Temple from his father! David's battles prepared for Solomon's building.

Everything you invest in the physical body you have will be gone before you are or buried with you when you die. **The significance of your life will be based on the value of your investment.** So, again, I ask you: How much of you will they bury?

If you only invest in the "here and now," you are in trouble. However, even the smallest investment in the next generation will be rewarded. When you invest in others, you will live beyond your lifetime! What will you leave behind when you leave this life?

◇◇◇

DAY 73 — HOW WISE ARE YOU?

I Kings 11:1,3b But king Solomon loved many strange women, together with the daughter of Pharaoh, women of the Moabites, Ammonites, Edomites, Zidonians, and Hittites...and his wives turned away his heart.

Are you as smart as you would like to be? Do you want to be any smarter and wiser than you are right now? Think about the wisest man to ever live, Solomon. How wise was he? He certainly ruled his kingdom wisely, but please remind me again of the number of wives Solomon had? How smart is that?!

Any man with great skill will ultimately fail if he has no wisdom in his private life. Skill and wisdom are not mutually exclusive, but one without the other is not smart. You may rule the world in great wisdom; but if you do not do right as a child of God in your private life, you are at a distinct disadvantage.

Without God's wisdom in the most basic areas of life, you cannot truly succeed. I thank God for the ministry He has given me, but **if I lead with great authority and wisdom yet fail to follow God's voice for me, I lose the game.**

I want wisdom. I've asked for it for the Bill Rice Ranch, my family, and the good folks we are privileged to serve. You need wisdom for your day, your job, and your family. If you will begin with wisdom in your private life, it will spread to all areas in which you have a stewardship.

LEADING AND SERVING

DAY 74

I Kings 12:7 And they spake unto him, saying, If thou wilt be a servant unto this people this day, and wilt serve them, and answer them, and speak good words to them, then they will be thy servants for ever.

I cannot think of a quicker way to ruin a kingdom than the way Solomon's son, Rehoboam, did it in I Kings 12. He was doomed from the beginning because he brought it upon himself!

The people came to King Rehoboam for leadership, along with sympathy for and relief from their burdens, but that is exactly what he did *not* give. **Serving is not in our DNA; being served is!** The problem is, you cannot serve others if you are self-serving.

If you are going to lead, you must lead yourself and serve others. Too often, we don't think of serving when we want to lead. **Any leader must discipline himself as well as serve those he leads.** Don't worry whether anyone will follow you. Lead yourself and serve others, and let God worry about the influence you will have.

You are not leading your ministry, your company, or your family if you are serving yourself. Don't allow "serving God" to become an alibi for "serving me." Beware the man who wants you to follow him when he follows no one else. May the people following your leadership today find one who is submissive to God and serving others.

DAY 75

HEAR OR HIDE?

I Kings 14:6 And it was so, when Ahijah heard the sound of her feet, as she came in at the door, that he said, Come in, thou wife of Jeroboam; why feignest thou thyself to be another? for I am sent to thee with heavy tidings.

Have you ever tried to "enhance" your résumé or accomplishments? Those who can't enhance it might try to hide it! That is exactly what a king tried to do when he needed God's help. Jeroboam thought his problem was the health of his son; God knew Jeroboam's problem was the health of his soul. Jeroboam had his wife disguise herself when she went to visit the prophet Ahijah. Well, Ahijah couldn't see very well (verse 4), so the disguise wasn't really needed! A prophet without eyes was going to help a king without sight, but God could see everything perfectly.

Jeroboam didn't want to hear what he needed to hear. He tried to hide reality from the prophet and from God, but does it make any sense to hide from the God of the Universe? The same is true with us. **God sees right through to our core**, so no amount of enhancing or manipulating will get past Him.

God perfectly and completely knows you as you are, and He speaks truth. You can either hear or you can muffle what God is saying to you today. God knows you and loves you enough to tell you what you need to hear. Don't disguise it or hide from it!

FED OR FED UP?

DAY 76

I Kings 19:4 But he himself went a day's journey into the wilderness, and came and sat down under a juniper tree: and he requested for himself that he might die; and said, It is enough; now, O LORD, take away my life; for I am not better than my fathers.

" That's enough!" Have you ever had a day like Elijah was having in I Kings 19? After he received some scathing threats from Jezebel, he sat down under a juniper tree and said, *"It is enough; now, O LORD, take away my life…."* You may have not felt *quite* that bad; but I must admit that some days, I have thought the life of a bird sounded pretty good. I could put up with a steady diet of worms for the chance to fly away!

Elijah was fed up and stressed out. To top it off, an angel woke him up and provided a meal for him because *"the journey is too great for thee."* Elijah was fed up with where he had been, and where he was going was more than he could handle. In his mind, he was overwhelmed and outnumbered!

If you find yourself discouraged like Elijah, don't forget the hand of God in your life. **Elijah was lonely, on the run, hungry, and hiding** *precisely because* **he was doing right!** God sustained Elijah in a number of ways; and He used birds, a widow woman, and an angel to feed him. Truthfully, God had led him, protected him, and fed him all the way, and He can do the same for you! You can't replicate receiving provision directly from the hand of God with anything else.

Like Elijah, you may be discouraged this morning while you are reading this. If not, your time is probably right around the corner! Whether you are discouraged by the wicked or by other believers, don't forget the hand of God in your life. When you are doing right but finding discouragement on every side, instead of getting fed up, remember from Whose hand you've been fed!

DAY 77

GOD OF ALL

I Kings 20:28 And there came a man of God, and spake unto the king of Israel, and said, Thus saith the LORD, Because the Syrians have said, The LORD is God of the hills, but he is not God of the valleys, therefore will I deliver all this great multitude into thine hand, and ye shall know that I am the LORD.

What kind of God do you have? Some people have a "health and wealth" god. Others don't even think about their god until something goes terribly wrong. Some just have a god for everything! They have a god for harvest, a god for drought, a god for battle, a god for the sun, and a god for rain. The Syrians no doubt had this type of mindset. When they kept losing battles to the Israelites, they figured a switch in strategy and location was needed. They were sure that the God of the mountains could not win in the valleys.

From this story we learn that **the only God that will do you any good is the God that is God of all.** The God we serve *is* the God of the mountains *and* the valleys—He is God of ALL. Wherever you are in life, God is there. If you find yourself in the "valley," God is there. If you are on a mountain today, God is there.

God is a gracious God. In I Kings 20, do you know the *"king of Israel"* mentioned in verse 28? God was actually helping the wretched King Ahab! The purpose of His graciousness towards Ahab was that he would *"know that I am the LORD."* God didn't help him because he was good; God did so because of His graciousness! The same is true with you. You have no goodness to offer God, but He has grace to offer you!

God is the LORD God. He is the eternally self-existent God of all creation. He is not an idea formed by our culture. He is the One who made us! No god you can make is worth serving.

The God we serve is the God who made everything! His power exceeds any problem we have.

Whether you are sunk in the depths of a valley, or you are riding high in the saddle on a mountain top, don't forget that God is the God of the valleys and the mountains. He is gracious and all-powerful. **He is the God of all!**

◇◇◇

LIVING FOR HERBS

DAY 78

I Kings 21:4 And Ahab came into his house heavy and displeased because of the word which Naboth the Jezreelite had spoken to him … And he laid him down upon his bed, and turned away his face, and would eat no bread.

King Ahab was a beggar that did not know how to be full. In I Kings 21, his priorities were completely out of whack! Ahab's neighbor, Naboth, had a vineyard that backed up to the king's palace. Ahab wanted it for an herb garden, and he made an offer of a replacement vineyard or the value of the vineyard. In the king's mind, there was nothing special about Naboth or the vineyard, but Naboth called it *"the inheritance of my fathers."* Jezebel then framed Naboth and had him killed—all for a silly vineyard!

What is important to you? What is your highest priority in life right now? You don't have to answer; it's obvious to everyone around you! People know your priorities by **what keeps you awake at night, what energizes you during the day, what you have passion for, what you will pay for,** and **what you pursue.** Ahab wanted herbs, so he offered money and eventually took a life! The Bible says of Ahab, *"But there was none like unto Ahab, which did sell himself to work wickedness in the sight of the LORD, whom Jezebel his wife stirred up."* **A person who pursues "low level" priorities sells himself cheaply.** Much of what consumes our days is low on the scale of importance. Sometimes herbs aren't worth the price you must pay!

Ahab sold his life cheaply because he pursued cheap things. He was a beggar in a king's palace! It is amazing how much energy and resources we can put into things that don't amount to much. Don't pursue after "herbs" when there is so much more to life. Perspective for your priorities will help you live for the highest priority. As Matthew 6:25 reminds us, *"Is not life more than meat, and the body more than raiment?"*

◇◇◇

DAY 79 — PRECEDENTS

I Kings 22:52 And he did evil in the sight of the LORD, and walked in the way of his father, and in the way of his mother, and in the way of Jeroboam the son of Nebat, who made Israel to sin:

Who was the most influential king of Israel? You could certainly make a case for David. You might even say Solomon. You could also make an argument for Ahab. The Bible says that Ahab *"did evil in the sight of the LORD above all that were before him."* On top of that, he married Jezebel and served Baal. I would want no part in any of that!

However, I really do believe the most influential king of Israel was Jeroboam! I think that because the Bible mentions him repeatedly. Twenty kings are mentioned covering a span of 210 years, and not one of them did well. All of them followed the precedent of one man: Jeroboam.

Someone who sets a precedent obviously *precedes*. A precedent setter sets the tempo. **There are really three kinds of people: precedent setters, precedent takers, and precedent breakers.** You may be more than one, but which describe you?

You will precede every generation that will ever be born hereafter. You are a **precedent setter** whether you realize it or not. The question is not *if* you will set a precedent; the question is, *what kind* of precedent will you set? What kind of "tempo" are you setting in your home?

Perhaps you are a **precedent taker**. That is, you are deciding if you will take the cue of those who preceded you. You should not mindlessly accept everything before you. Use God's Word and God's Spirit to decide what is right and worthy of accepting. There is something to be gained by taking wise and godly precedents as your own. What precedents should you take as your own?

A **precedent breaker** does not accept the status quo for life just because "that's the way it's always been." Do not buy the lie that just because someone did, you must. Don't buy their script for your life! Remember that it is wise to learn from the mistakes of wise men; it is foolish to replicate their mistakes.

You are setting a precedent right now by your attitude, actions, and thoughts. You have some precedents in your life that need to be taken as your own. Some precedents set by others before you might need to be broken. By God's grace, we ought to be wise precedent setters, takers, and breakers!

EMPTY VESSELS

DAY 80

II Kings 4:3 Then he said, Go, borrow thee vessels abroad of all thy neighbors, even empty vessels; borrow not a few.

Do you have what it will take for today? No matter your answer, I am sure you know the feeling of *not* having what you need! That is why I love the story of the widow in II Kings 4. She did not have what she needed for the week; but as she obeyed God, she had everything she needed for the day.

God fills what is empty and what is ready to receive. If you're not empty, He can't fill you; and if you're not eager and looking for it, He can't fill you either! This widow was certainly empty. She said, *"Thine handmaid hath not any thing in the house, save a pot of oil."* Isn't it strange, then, that Elisha told her to *"Go,* *borrow thee vessels abroad of all thy neighbors, even empty vessels; borrow not a few."* Borrow vessels for what? She didn't have anything! However, she "went for broke," borrowing anything empty and available from her neighbors!

Amazingly, when she finished pouring her oil into the last borrowed vessel, *"the oil stayed."* What she had was sufficient because she was empty! The pots she borrowed were empty vessels. She may not have had what was needed for the week, but she had what she needed for the day. **You don't need everything for the entire week; you live life one *day* at a time.** If you will use what you have today, you will have what you need next week. Prepare for tomorrow, but live in today.

God fills when we are ready to receive. This widow reminds us that we are ready to receive when we are empty and eager. What "empty vessels" do you have that God can fill?

◇◇◇

DAY 81

"FEAR NOT"

II Kings 6:16 And he answered, Fear not: for they that be with us are more than they that be with them.

I love the story in II Kings 6! The Syrian king sent out a posse to round up Elisha. When Elisha's servant woke up that morning, he saw a whole army surrounding the city. His response was, *"Alas, my master! how shall we do?"* Elisha's answer was profoundly simple: *"Fear not."* He also prayed that the servant would be able to see and comprehend what Elisha already knew. When the servant's eyes were opened, *"he saw …the mountain was full of horses and chariots of fire round about Elisha."* They were surrounded by God's heavenly army!

Elisha's answer to his servant is God's answer to you today: *"Fear not."* Perhaps you are tempted to fear because you are outmanned. Elisha and his servant were completely surrounded! Maybe you feel troubled because you feel outmaneuvered. The enemy had the high ground and completely surrounded Elisha. In short, Elisha and his servant were completely vulnerable! Do you feel that way today? There will always be reasons to fear; God gives us a reason *not* to fear.

The lesson to be learned is that there is more to your day than meets your eye. There is more to your battle than you can see! Romans 8:31 reminds us that *"if God be for us, who can be against us?"* It does not matter how large your enemy is; it matters how big your God is. We can spend so much time calculating the size of our enemy that we totally forget the power of God. Whatever your day holds, there is more than will meet your eye. God's answer for the trouble that surrounds you is, *"Fear not."*

◇◇◇

TREASON! TREASON!

DAY 82

II Kings 11:14b ...and Athaliah rent her clothes, and cried, Treason, Treason.

Nobody likes a traitor; everybody thinks loyalty is a virtue. But the real question is **to whom am I loyal?** In II Kings 11, we read of a power-hungry queen named Athaliah who *"destroyed all the seed royal,"* which meant killing off her own grandchildren and family. Without her knowledge, her grandson Joash (the rightful heir to the throne) was saved from the murderous rampage. When he was later brought out from hiding and crowned king, Athaliah cried, "Treason, Treason."

What a twisted story! The eleventh chapter of Second Kings is all about loyalty and treason. Who was loyal? Who was treasonous? Were the ones who sheltered the king-to-be committing treason, or was the woman who was crying "treason" the one?

Sometimes we elevate loyalty to a virtue without considering to whom we are being loyal. **Any loyalty that hinders your obedience to God is a treasonous act to be avoided.** I have to be careful about this truth, and I suspect you do too. It is natural to concern yourself with others' loyalty toward you instead of your loyalty to God. The moment I stop following God, I forfeit any right to expect loyalty from anyone else.

Loyalty is important, but the way to gain it is to pursue an authentic loyalty to God. Any loyalty you receive from those under your authority ought to be channeled to the One who gave you the position. Any loyalty you hold that hinders obedience to God is treason and should be avoided. Are you loyal? Are you a traitor?

DAY 83 — CUT DOWN TO SIZE

II Kings 14:11 But Amaziah would not hear. Therefore Jehoash king of Israel went up; and he and Amaziah king of Judah looked one another in the face at Bethshemesh, which belongeth to Judah.

Recently, someone on the Ranch tried to teach my boys the art of trash-talking. I guess he thought they needed help in this area! It was humorous near the end of the week when the teacher became the focus of the good-natured jesting. Professional sports are full of trash-talking players, and we find a sort of "trash-talker" in II Kings 14.

Amaziah was feeling good about himself and the strength of his army, so he challenged Jehoash to a fight. Jehoash used an analogy to cut Amaziah down to size, calling him *"the thistle that was in Lebanon"* while calling himself *"the cedar that was in Lebanon."* This little thistle thought he was big and tough like the cedar tree until a passing animal accidentally squashed the lowly thistle. Jehoash said in verse 10, *"Thou hast indeed smitten Edom, and thine heart hath lifted thee up … why shouldest thou meddle to thy hurt, that thou shouldest fall, even thou, and Judah with thee."* And that is exactly what happened! Amaziah got his face-to-face battle, but he and his country were *"put to the worse before Israel."*

This story illustrates an important lesson for us today: **don't wait for someone else to cut you down to size!** Do it yourself before you get in trouble. **You are due for defeat when you cannot see**

yourself honestly and will not hear reproof. Amaziah couldn't see who he truly was, and he would not listen. He thought he was a cedar tree when he was only a thistle! He was "a big fish in a small pond."

The standard by which you judge yourself is important. If you judge yourself by other people, your estimation will always be misguided. This sounds simple, but there are more people in the world than your little circle of friends and peers. A wrong view of self and a failure to listen will always lead you astray!

The best way to guard against the folly of Amaziah is to see yourself as you are and to listen. Don't judge yourself by another person—you can always make yourself seem good if you narrow the field enough! A humble spirit, a correct view of self, and a willingness to hear will keep you from needing to be cut down to size!

ACTIVE AND PASSIVE

DAY 84

II Kings 15:9 And he did that which was evil in the sight of the LORD, as his fathers had done: he departed not from the sins of Jeroboam the son of Nebat, who made Israel to sin.

It is a sobering thought to realize that God summarizes the lives of the majority of the people you read about in the Bible in one verse. Many times we fly right through the passages without thinking about them at all! The vast majority of people, even kings, only get one verse to cover their entire lives.

II Kings 15 says the same thing about four kings (verses 9, 18, 24, 28): *"And he did that which was evil in the sight of the LORD … he departed not from the sins of Jeroboam…."* God's epitaph for each life was exactly the same! They did evil in the sight of the Lord by not departing from the sins of Jeroboam. What Jeroboam did by design, four generations of kings did by default.

Jeroboam followed his sin on purpose; those after him followed his sin by following precedent.

So many times, we do what we do because of precedent. We take the "script" someone has given us and blindly follow it without taking thought. That's not half smart! **What one man does actively, others will follow passively.** Centuries of kings did wrong by simply being passive and doing nothing.

Sometimes to do nothing is to do wrong. In fact, the easiest wrong to do today is to do nothing! Remember that your actions and decisions today will chart the course for people you will never see or meet. Jeroboam never saw King Zachariah, King Menahem, King Pekahiah, or King Pekah, but they surely followed the pattern he had set. Be purposeful in what you do today. What you do actively today will be done passively tomorrow.

◇◇◇

DAY 85 — THE WAY TO MOVE EARTH

II Kings 19:14 And Hezekiah received the letter of the hand of the messengers, and read it: and Hezekiah went up into the house of the LORD, and spread it before the LORD.

What is the worst day you can think of? Do you have some specific ideas, maybe from personal experience? Hezekiah was having one of those kinds of days in II Kings 19. The most powerful man in Israel found himself with no power. On top of that, he was standing against the most powerful army on earth. So what did Hezekiah do?

Verses 14-15 tell us that Hezekiah took the threatening letter from his enemy Sennacherib and *"spread it before the LORD"*—he prayed about it! Verse 3 says about that day, *"This day is a day of trouble, and of rebuke, and of blasphemy...."* The situation was truly dire, the need was great, and Hezekiah literally spread his problems before the Lord. What a wonderful description of prayer!

Have you ever heard the phrase "moving heaven and earth"? The phrase means someone really shakes things up and gets things done. **But the way you move earth is by moving heaven; you have power on earth with power from heaven.** Do you ever find yourself in a place like Hezekiah? You can try to build yourself up and gain support by working in your flesh to "move Earth," or you can ask God for His power from heaven. Evangelist Jim Cook has said, **"God is moved by our faith, not our work."** *Who* is helping us do our work is the key!

When you are in a place like Hezekiah's, **you will pray with more boldness when you make God's cause your own.** God had the power that Hezekiah needed, and he clearly asked for it. He boldly and reverently put it before the Lord in verse 17: *"Of a truth, LORD, the kings of Assyria…."* When God's cause is your cause, you can pray with even more boldness!

Perhaps you have specific problems that are on your mind today. It might do you well to actually write them down on paper. Now, Hezekiah didn't write out his prayer, but he did write out his problems. Next time you are faced with worries and problems, try writing them down and spreading them before the Lord. There is nothing magical about that; but if nothing else, it will help you be clear and specific instead of vague!

"And Hezekiah went up into the house of the LORD, and spread it before the LORD."

DAY 86

HOW TO BE HEARD

II Kings 22:18-19 ... As touching the words which thou hast heard; Because thine heart was tender, and thou hast humbled thyself before the LORD, when thou heardest what I spake ... I also have heard thee, saith the LORD.

Perhaps this scenario has happened to you: you dial a phone number, the line rings, a voice answers, and either a) you think it is a voicemail recording when it is actually a person or b) you think you are speaking to a person when, in reality, it's a recording. I can't tell you how often I've called a church or cell phone and really thought I was speaking to the voice I heard, only to quickly find out that I was having a conversation with myself! You can't have a two-way conversation with someone's voicemail!

Instinctively, we want to be heard. We want our prayers to be heard by God. King Josiah reminds us in II Kings 22 that **to be heard by God, we must hear God.** We have a wonderful advantage over believers in the Bible; we have the complete Word of God. Josiah had to go to a central place where a single copy of God's Word was kept. Most of us only have to go a few feet in our house to find one of the many copies of the Bible we own! So we must take time to listen to God's Word to us.

If we are going to hear God, we must have humility of heart. Verse 19 says, *"Because thine heart was tender, and thou hast humbled thyself before the LORD...."* **So many times we are hard of heart because we are hard of hearing!** This sounds simple, but you cannot hear God if you are not reading His Word. Reading the Bible takes time and is important, but there is nothing spiritual about reading a passage yet missing the point of what you are reading. We all could probably use much more thinking (the Bible word is *meditation*) in our daily Bible reading.

You will be amazed what God has to say to you in His Word when you take time to actually listen! Humble yourself and hear what He is saying to you. **To be heard, you must hear God.** In the busyness of life, don't forget to hear from God!

FOLLOWING THE SCRIPT

DAY 87

II Kings 23:24 Moreover the workers with familiar spirits, and the wizards, and the images, and the idols, and all the abominations that were spied in the land of Judah and in Jerusalem, did Josiah put away, that he might perform the words of the law which were written in the book that Hilkiah the priest found in the house of the LORD.

In the summer of 2013, Nik Wallenda crossed the Grand Canyon on a tightrope. He walked on a 2-inch thick steel cable suspended 1,500 feet above the Little Colorado River. While this daredevil attempt seems insane (I love hiking the Canyon, just not on a tightrope!), it is the type of script the Wallenda family has followed for generations.

I wonder what your script is. "The Rices (or your family name) have always done _____." Maybe your family has done the same thing for decades. Sometimes a "script" can be helpful, but **you must decide what script you are going to follow**. Josiah was a man with centuries of pagan, godless, idol-worshipping relatives, but he went off script and served the Lord. He didn't follow the script; he followed Scrpture. Josiah was a king of action!

Will you follow the script or the Scriptures? There is no reason to repeat the actions of past generations that do not line up with God's Word! We are living in a day when anything the Bible says is completely counter to our culture. Pop psychologists are not the authority; Hollywood isn't the authority; and my great-great-grandfather isn't the authority, either. **God is *the* Authority.**

What Josiah did was radical, but only because wickedness had become the standard. He broke the precedent that Jeroboam and many kings before him had set. So, in your life, which script will you follow? Which authority will you follow?

Following Scripture will enable you to see farther, clearer, and deeper than you otherwise would. See life through the eyes of

God by reading His Word and following His script for you. You are not a slave to the script you've been handed!

DAY 88

THE STATUS QUO

II Kings 24:14 And he carried away all Jerusalem, and all the princes, and all the mighty men of valour, even ten thousand captives, and all the craftsmen and smiths: none remained, save the poorest sort of the people of the land.

We have witnessed the destruction of a nation. We have front row seats to a nation that rejects God, sheds innocent blood, and follows idolatry. It may surprise you that I am not talking about the United States of America. I am talking about the nation of Judah, God's chosen people.

Judah wasn't destroyed by God; Judah destroyed herself. The Bible records her conquest and destruction at the hands of Babylon in II Kings 24-25. Interestingly enough, tyrants have used the same methods from the very beginning! If you were poor, dumb, or had no fight in you, you were left behind. **The *status quo* was good enough for the enemy when it posed no threat.** The skilled, smart, and strong were taken captive by Babylon.

If you are willing to fight for what's right, you can certainly expect a fight to come. Dads, if you are willing to put up a fight for what you believe, you can expect one. Moms, if you are willing to put up a fight for what you believe, you can expect one. Teenager, if you are willing to put up a fight for what you believe, you can expect one.

If you want to live the "easy" Christian life, the Devil will be your best ally. If a teenager wants to live for popularity and influence, the Devil is happy to oblige. If a dad wants to gain the world, the Devil will be happy to help that dad lose his own soul. As

Evangelist Jim Cook says, **"The devil will give you what you want, but it will cost you what you have!"**

You do not have to seek out trouble or confrontation; you will have plenty just by standing for what is right. If you are not a threat to the Devil, he does not have to be concerned with you. However, standing for truth and right will certainly bring a flood of attacks and opposition. Staying in line with truth is never an accident; it is an intentional, costly decision. By God's grace, determine that you are going to challenge the *status quo*!

WHAT'S IN A NAME?

DAY 89

I Chronicles 4:9-10a And Jabez was more honourable than his brethren: and his mother called his name Jabez, saying, Because I bare him with sorrow. And Jabez called on the God of Israel....

What do you think of when you hear the name *Jabez*? If you know anything about Jabez, you probably think of his prayer in I Chronicles 4. More than likely, you do not think about what his name really means: sorrow! Why did his mother name him "sorrow"? I'm not sure, but it was not exactly a smooth beginning to his life!

Years ago when my grandfather went to the Congo to conduct revival meetings, it was a common practice among the tribal people to name their kids the worst name they could think of. In their thinking, the Devil hated them and would do their kids harm if they appeared to love them. Some chose names like "Utterly Worthless" and "Ugly Monkey"! Can you imagine getting a name like that or a name like "sorrow"?

Jabez *"called on the God of Israel,"* but more importantly, he saw the "God of Israel" as the God of Jabez. In contrast to David's sons, Jabez had no privileged standing at birth; his own mother named him sorrow! Don't miss the truth that you can do one of two things with the name you have today: you can accept it blindly, or you can do something about it. **What your name** means to other people is purely a matter of what you make it mean!

Today, you have the name your parents gave you and the name you give yourself. Jabez started out as sorrow, but he ended up with God giving him what he asked for! A dishonorable name by birth was transformed into an honorable name by the way he lived. Begin today with the end in mind. That is, if you want to end up being reliable, start that way today. If you want to end up being honest, determine to be truthful in everything today. What's in a name? Well, I guess that depends on what you make of it!

◇◇◇

DAY 90

CALLED TO BE ... A JANITOR?

I Chronicles 9:22 All these which were chosen to be porters in the gates were two hundred and twelve. These were reckoned by their genealogy in their villages, whom David and Samuel the seer did ordain in their set office.

James Porter Rice was my great-great grandfather, so the name Porter was one my wife and I were considering putting on our newborn son. Names mean something; and when we looked up *Porter*, we found that in Hebrew, it means "janitor" or "gatekeeper." You know, janitor is not a very exciting name for a person or a job, unless you are the janitor for God!

In chapter 9, we read about a group of select people who were janitors (gatekeepers). Just look at the words used to describe them: *chosen, ordain, set office, oversight,* and *charge.* Many of us probably have never aspired to be a janitor or gatekeeper. But everything changes when you are serving God Almighty! There is an entire chapter in the Bible about janitors!

The One you serve determines the worth of your service. That means if you are self-serving today, your day will not be worth much. If you are serving some boss or supervisor, the worth of your work will be minimal. Even serving a good ministry will put a cap on your worth and work. However, if you are serving

God today, it will not matter what you are doing specifically. Even serving as a janitor is a high calling if the One you are serving is God!

Interestingly, we are told that some of these porters were descendants of Korah. Yes, *that* Korah! In Numbers 16, Korah was upset because he did not like his job and was jealous of the job that God gave to Moses and Aaron. Moses said to him, *"Seemeth it but a small thing unto you … to do the service of the tabernacle of the LORD … ?"* (Numbers 16:8) What God had chosen and ordained for Korah wasn't good enough for him! I don't know what God has made for you to do today, but **you can't do better than what He has chosen for you!** Whether you are a janitor, a doctor, or a preacher, remember that the One you serve determines the worth of your service.

THE TEST OF ADVICE

DAY 91

I Chronicles 13:3 And let us bring again the ark of our God to us: for we inquired not at it in the days of Saul.

Is it possible to get bad advice from competent people? David did not know what to do with the Ark of God. Although his heart was in the right place, he asked advice from everyone under the sun except for the One Who shined on them. He said in verse 2, *"Let us send abroad unto our brethren every where … also to the priests and Levites…."* Everyone was in agreement about it, and as verse 4 puts it, *"For the thing was right in the eyes of all the people."*

Did David ask advice? Yes. It even sounds noble and caring that he asked everyone what they thought. Although he asked for advice, did David receive the right answer? No! This resulted in the death of a man named Uzza, because he touched the ark. Uzza tried to steady the ark on the *"new cart"* they were using to transport it. He meant well, but meaning well and doing well are two different things.

David knew what he had done and what was needed in verse 12: *"And David was afraid of God that day, saying, How shall I bring the ark of God home to me?"* He had asked everyone except God. It was not wrong to ask advice—we *should* ask advice. But David failed to ask the very One Who knew the right answer!

The test of consultation is whether or not you are heading in the right direction after you receive the counsel. The purpose of advice should be to get the right answer. Others may give you wrong advice, or you may be asking for some self-serving purpose. However, when asking advice, remember to ask the One Who knows the right answer ten times out of ten.

◇◇◇

DAY 92
YOUR GREATEST DANGER

I Chronicles 15:26 And it came to pass, when God helped the Levites that bare the ark of the covenant of the LORD, that they offered seven bullocks and seven rams.

At first glance, I Chronicles 15:26 does not look like an important verse; but I guarantee you that it was important to the Levites! The Bible says that *"God helped the Levites that bare the ark…."* The word *help* means "to surround, defend." God helped the Levites in a specific way—He surrounded them and protected them. From what or whom did the Levites need protection? Well, think about Uzza just a few chapters before! Would *you* want to move the ark when the last guy died doing the same thing? They needed protection from the judgment of God!

God can either be your greatest danger or your greatest defense. Israel's greatest danger in bringing the ark back was not the advice of friends or the assault of enemies. Their greatest danger was God! When they brought the ark back, there was a sense of thankfulness and praise.

So what is the upshot of all of this? Whatever you've worried about today, you have probably not worried about God. So much of our days are consumed with pressure from a boss, from coworkers, or from peers. Your entire day can be shaped by what you think about what people think about you! Instead of being consumed with what people think, be concerned with what God already knows about you. Your greatest danger is not displeasing someone at work or school; **your greatest danger is in forgetting God!**

There is no need to spend your day under a cloud. Frame all of your cares in light of who God is and what He can do. Is God your greatest danger, or is He your greatest defense?

◇◇◇

YOUR DREAM JOB

DAY 93

I Chronicles 18:8 Likewise from Tibhath, and from Chun, cities of Hadarezer, brought David very much brass, wherewith Solomon made the brasen sea, and the pillars, and the vessels of brass.

Back when I was fifteen and working in the dining hall at the Bill Rice Ranch, I thought my boss was giving me my dream job. He asked a group of us guys, "Who knows how to drive a stick shift?" That is a magic question to a bunch of high school guys! I had my dream job in my head before he finished his question. Whatever it involved, if it involved a stick shift, an engine, and four tires, I was in! Instantly, five or six hands shot up in the air. My boss picked the six guys with hands raised … and promptly gave us each a broom to start sweeping! It did not take us long to figure out that we had been hoodwinked. Sweeping was not exactly the dream job I had in mind.

Have you ever gotten a job, but not the one you wanted? Well, even kings don't always get their dream jobs! King David wanted to build God a house, but God did not give him that privilege. Instead, God passed the privilege of building the Temple to

David's son, Solomon. So what do you do when you do not get your dream job? **David's example challenges us to do what we *should* do, even when we cannot do what we *want* to do.**

The first part of chapter 18 is a record of David's work, even though he could not build the temple. He *smote, took, houghed, reserved,* and *brought.* Those are all words of action. What came of all of this? Solomon built with what David had brought. When David could not do what he wanted to do (build the Temple), he did what God wanted him to do. David's battles paved the way for Solomon's building.

You may not be doing what you want to do, but you can always do what you should do. **Do what you should do, regardless of who reaps the benefits.** Solomon reaped the benefits of the battles David fought and the materials David gathered. **Do what you should do, regardless of who gets the credit.** When you think of the magnificent Temple in Jerusalem, whose name do you associate with it? We know it as *Solomon's* Temple. Solomon's Temple was David's idea, not Solomon's!

You might be disappointed in a job you get today. It may not be your dream job, but you can learn from David's example. When you cannot do what you want to do, do what you should do, no matter who reaps the benefits and who gets the credit.

◇◇◇

DAY 94

GOD'S PATTERN

I Chronicles 28:19 All this, said David, the LORD made me understand in writing by his hand upon me, even all the works of this pattern.

I'm not a big fan of following the directions that come with bicycles, phones, or other presents at Christmas, but I've found that following the directions is an important part of doing things correctly. David had a magnificent temple that he wanted to build, and God gave him the pattern for it.

Patterns are important. Directions for work, school, and even a phone are important, even though the directions themselves may not be glamorous or exciting. **Your work will be no better than the pattern you follow.** Taking the time to read the instructions is paramount.

God has given us a pattern for life in His Word. He knows you, and He can see the "big picture." You cannot afford to miss reading God's instruction manual for life—the Bible. Nothing good you will do today can replace the need you have to follow God's pattern. And you'll never know His pattern until you read It! Even taking a bite-sized portion each day and thinking on that passage throughout the day will make a huge difference. At times, I've taken a 3x5 card and written a verse on it to help remind me of God's truth throughout the day.

God's pattern provided a reason for all the gifts David gave to Solomon. Why did David give gold and silver (verses 14-18)? Because God's pattern called for them! **God knows you, and He knows best.** That is why we ought be concerned with reading His pattern for our lives in the Bible. Don't live today without knowing what the Maker calls for in the pattern! Your work will be no better than the pattern you follow!

A NEGLECTED CONSCIENCE

DAY 95

II Chronicles 8:11 And Solomon brought up the daughter of Pharoah out of the city of David unto the house that he had built for her: for he said, My wife shall not dwell in the house of David king of Israel, because the places are holy, whereunto the ark of the LORD hath come.

What comes to mind when you think of King Solomon? Wisdom. Wives. Wealth. (That makes an alliterated three-point sermon right there!) Just think about those three for a minute. You would think a man with loads of wisdom would be wise enough not to have seven hundred wives! Seriously, isn't that at least a little confusing?

Solomon neglected his conscience. The Bible says that Solomon built a separate house for Pharoah's daughter because *"the places are holy, whereunto the ark of the LORD hath come."* **Solomon's conscience was obviously troubled by this dilemma.** If you feel convicted that a holy God will be dishonored by what you are doing, then you shouldn't be doing it!

Do you think some politicians have two lives? Yes, of course! Just watch the news! It is easy to see the wickedness in a politician's secret personal life that contradicts his public reputation. What is not so easy is to see that when duplicity is in *my* own life. It is amazing the number of things we scorn in the lives of politicians that we allow in our own lives.

Solomon **responded to his conscience superficially** and **underestimated the danger of neglecting his conscience.** How did he get to the place where he had seven hundred wives, including a pagan daughter of Pharaoh? It all started with neglecting his conscience the *first* time. That is the way sin works; a person does not *fall*— he *slides*. A neglected conscience in one area starts the slide into sin. Neglecting your conscience always leads you in the wrong direction, no matter how far down the path you are!

Ask the Lord for determination with your conscience. Don't let even one small area of conscience go neglected today! A tender conscience informed by God's Word and His Spirit will be invaluable as you go throughout this week. Whatever you do, don't neglect your conscience today!

PREPARATION

DAY 96

II Chronicles 12:14 And he did evil, because he prepared not his heart to seek the LORD.

How long did it take you to get ready this morning? You may not have all the time you want or need, but how long does it normally take you? Some people amaze me in how quickly they can get themselves prepared in the morning! Preparation is important, isn't it? How you appear and smell when your alarm goes off is, hopefully, not how you look and smell right now!

Preparation is important in other areas of life, as well. Do you like eggs? Do you like *raw* eggs? The difference is preparation! II Chronicles 12:14 reminds us that preparation, or the lack thereof, is critical to our spiritual lives. King Rehoboam *"prepared not his heart to seek the LORD."* He wasn't evil so much because of what he *did* do as he was evil for what he *did not* do.

Do you think Rehoboam ever prepared for anything? Of course! Verse 1 tells us that when Rehoboam *"had established the kingdom, and had strengthened himself, he forsook the law of the LORD, and all Israel with him."* He prepared (*"established"*) his kingdom and strengthened himself, but he did not prepare for God. Rehoboam prepared, but he didn't prepare for things that were most important.

Your preparations reveal your priorities. Are you strengthening yourself and your position, or are you seeking the Lord? Rehoboam did not prepare his heart to seek the Lord, and the Bible says that *"he did evil."* What are you doing this morning to prepare your heart today? With the Lord's help, prepare your heart and prioritize your day to seek and serve Him.

DAY 97

BEWARE AND BE STRONG

II Chronicles 15:2 And he went out to meet Asa, and said unto him, Hear ye me, Asa, and all Judah and Benjamin; the LORD is with you, while ye be with him; and if ye seek him, he will be found of you; but if ye forsake him, he will forsake you.

After defeating a massive Ethiopian army, King Asa was riding tall in the saddle! Once they gathered all kinds of spoil from the battle, God warned His people to beware and encouraged them to be strong. On the heels of great success often comes the strongest temptation to forsake God or faint in the way. The two admonitions from God to Asa and Judah are good for us today, too!

First, **beware when you are "riding high."** There is a danger of forsaking God because we do not usually seek God in times of peace. We are most likely to seek God when our backs are against the wall and we have no other option! When things are going well, beware! Seek God!

Second, **be strong when you must battle.** God encourages Asa and the people to *"be ye strong therefore, and let not your hands be weak: for your work shall be rewarded."* Are your hands ever weak? Do you ever face a "down" time? Serving God takes work, and sometimes that work can drain you—and then ask you for more. When you find yourself stretched thin, remember this verse: *"Your work shall be rewarded."* As assuredly as God will be found if you seek Him, your work *will* be rewarded. And the great news is that God is the One doing the rewarding!

My natural inclination is to wonder *how* God is going to reward me. How will this situation pan out? Crowns in heaven? Money now? Affirmation now? Keep in mind the truth that **the One who rewards you is more important than the rewards He gives!**

You do not need to know how God is going to reward you in order to trust Him. *"Be ye strong therefore, and let not your hands be weak: for your work shall be rewarded."* God will not forget to reward you! Today, if things are going great, beware lest you forsake God. If you are weary in well-doing, be strong and trust God for your reward!

"WHICH COULD NOT DELIVER"

DAY 98

II Chronicles 25:15 Wherefore the anger of the LORD was kindled against Amaziah, and he sent unto him a prophet, which said unto him, Why hast thou sought after the gods of the people, which could not deliver their own people out of thine hand?

It is frustrating to buy into something that does not deliver. Perhaps you buy the newest truck model with all the bells and whistles, but you find it cannot pull the heavy load that you need. Some domestic jobs can be just as frustrating when the equipment does not "deliver." No item in my home can frustrate me more than a vacuum that doesn't pick up dirt or an iron that doesn't iron.

In much the same way, thousands of people today will seek after gods that do not deliver. This amazing story of King Amaziah in II Chronicles 25 defies all logic. He defeated the Edomites, but then *"he brought the gods of the children of Seir, and set them up to be his gods, and bowed down himself before them, and burned incense unto them."* (verse 14) Amaziah knew God (verse 2), but he stole the gods of the people he just conquered and made them *his* gods! Let me ask you a few questions. Did the gods he stole hear the people that prayed to them? Did the gods care about the people? Did the gods deliver them from the armies of Judah? The answer to all of these questions is, no!

Instead of showing how great your God is, you can, like Amaziah, adopt gods that cannot deliver. One example is the "god" of money. You might hear yourself think or say, "If I only had this thing, then I would…." Most people spin their wheels by living their lives seeking a god that does not deliver. You cannot see God when you are serving other gods. In contrast, **God is**

a God that delivers. He has always delivered for the Bill Rice Ranch because He *can* deliver.

God sent a prophet to confront Amaziah, but he would not hear the prophet. The king said, "Forbear," which means, "Be quiet." After that, Amaziah never heard from God again. Verse 20 says, *"But Amaziah would not hear; for it came of God, that he might deliver them into the hand of their enemies, because they sought after the gods of Edom."* In the end, he lost what little he had (verse 24) because he did not follow God.

Do you know God? Are you seeking fulfillment from little "gods" that cannot deliver? May the example of Amaziah be a reminder to us today of the gods which cannot deliver.

◇◇◇

DAY 99
SUCCESS AND PROSPERITY

II Chronicles 26:5 *And he sought God in the days of Zechariah, who had understanding in the visions of God: and as long as he sought the LORD, God made him to prosper.*

Think about your greatest failure this year. After that failure, what was the one thing you wanted? Money? Wisdom? Time? Forgiveness? In one word, you wanted *help*.

Now think about your greatest success this year. After this success, what did you want then? In one word, you wanted *praise*. You wanted someone to notice!

When you are weak, you want help. When you are great, you want praise. So then, in which situation do you think you are in better shape spiritually? The trouble with prosperity is that you have a hard time seeking God. On the other hand, **the only way to prosper in God's sight is to seek Him!**

The Bible says of King Uzziah, *"As long as he sought the LORD, God made him to prosper."* Prosper is exactly what Uzziah did. Or rather, what God did for him. God *"helped him against the Philistines"* and Uzziah enjoyed fame, success, and strength.

He was living at the very pinnacle of success, and II Chronicles 26:15 records, *"For he was marvelously helped, till he was strong."*

Why was Uzziah strong? He was strong because God helped him. But there came a day in the Uzziah's life when he went, in his mind, from being helped to "helping God." We can be the same way! The result was this: *"When he was strong, his heart was lifted up to his destruction."*

Uzziah was not helping God; God was helping him! The day you forget the same is the day you are sunk. It would be better to be down and need help than to be on the pinnacle and want praise.

You can prosper, but only as you seek the Lord. Our nature is to stop seeking the Lord the moment He prospers us. **The sweet spot in life is to seek God, to prosper, and to continue seeking God.** Is that true of your life?

A WORLD WITHOUT YOU?

DAY 100

II Chronicles 32:31 Howbeit in the business of the ambassadors of the princes of Babylon, who sent unto him to enquire of the wonder that was done in the land, God left him, to try him, that he might know all that was in his heart.

Years ago, Jimmy Stewart starred in a motion picture called *It's a Wonderful Life*. (The movie has become somewhat of a tradition during the Christmas season.) The main character has experienced hard times and is thinking about taking his own life. He is given the opportunity to see what life would be like without him. Have you ever imagined what the world would be like without you? Think of all the things that would be different!

As different as this world would be without you, think of your world without God. That would be true misery! II Chronicles 32:31 says that *"God left him, to try him, that he might know all that was in his heart."* This particular time in Hezekiah's life is explained to us in II Kings 20. While his storehouses were full of gold and silver, Hezekiah's kingdom had been threatened by the enemies of God, and his own health was threatened by a serious illness. Hezekiah's story reminds us that God does not dwell with the rich and powerful; **He dwells with the needy and humble.**

It is true that this life is wonderful and that the world would be different without me. However, the world was here before me, and it will be here after me. This morning, the tragedy would not be a world without you or me; **the real tragedy would be a life lived without the God who made the world.** When, like Hezekiah, our storehouses are full, our health is good, and we have victory over our enemies, we are perhaps in our worst position.

This morning, you can do without many things, but you cannot do without God. Everything you may need this morning is provided by God Himself. Do you know God and His presence in your life? Whatever you may live without today, don't live without God!

◇◇◇

DAY 101

THE WORST THING

II Chronicles 33:12-13 *And when he was in affliction, he besought the LORD his God, and humbled himself greatly before the God of his fathers, And prayed unto him: and he was intreated of him, and heard his supplication, and brought him again to Jerusalem into his kingdom. Then Manasseh knew that the LORD he was God.*

What is the worst thing that could ever happen to you? Whatever your answer, consider this statement: the worst thing that could happen to you is to have nothing "bad" happen to you. Would you still seek God? Imagine that! Sadly, if you could live without ever needing God, you would—and so would I. That is the nature of our flesh, always bent toward wrong. **However, God loves us enough to not allow us to live without Him.**

King Manasseh was an evil king, so it is no surprise that evil came back to him. It was, no doubt, shocking when Manasseh *"besought the LORD his God, and humbled himself greatly before the God of his fathers, and prayed unto him...."* You would not have expected this at all!

Notice that Manasseh returned to God *"when he was in affliction."* Some people have this idea that it is wicked and selfish to pray only when you are in trouble. Can I tell you something? You are *always* in trouble! Sometimes you are just not smart enough to know that you're in trouble. God, in His goodness, allows our lives to inherit trouble to show us that we are in desperate need of Him. That is exactly why God *"brought upon them the captains of the host of the king of Assyria"* when they would not hearken. God brought the affliction because He loved His people!

The worst thing that could happen to you is to have nothing bad to happen to you. You do not ask God for help because you are good; you ask Him for help because you are needy. And if a king can humble himself, you can too.

God is known both by His power and by the might of His mercy. Mercy is not deserved or earned by the person receiving it—that's why it's mercy! God has power to act (sending Assyria against Manasseh) and to *not* act (showing mercy to Manasseh). No one else can give the kind of mercy that God can give. Today, you can live in confidence, knowing that **both the good and the bad things in life are conspiring to help you know that you need God.**

DAY 102

LOSING THE BOOK

II Chronicles 34:15 And Hilkiah answered and said to Shaphan the scribe, I have found the book of the law in the house of the LORD. And Hilkiah delivered the book to Shaphan.

Did you ever dream about being royalty when you were a kid? I don't know if Josiah ever dreamed about it, but he was **eight** years old when he was made king! Can you imagine an eight-year-old ruling a country?! What we do know about Josiah is that he *"did that which as right in the sight of the LORD"* and *"in the eighth year of his reign, while he was yet young, he began to seek after the God of David his father."* (verses 2-3)

One of his first orders of business in verse 3 was to *"purge Judah and Jerusalem from the high places, and the groves, and the carved images, and the molten images."* After he purged out the wickedness, he began to repair the house of God (verse 8). While repairing the house of God, Hilkiah, the priest, found a book of the law.

Why did they lose the book in the first place? They disregarded God's Word because they lost their heart for God. Consequently, they lost what tied them to God, as well (the book of the law). So how did they find the book? When they had a heart for God, they found it because they were looking for it! They were seeking God.

There is a great danger of "losing the book" in our everyday lives also. Sometimes the danger comes in "not having time"; other times we think we "already got it" because we go to church three times a week, attend special revival meetings, or even work in a ministry where we are surrounded by the Book on a daily basis. Still yet, we can simply be content to hear about Bible truth instead of seeing it for ourselves.

Reading the Bible is a very practical thing. You have to give It time—you need to take time. Some books of the Bible require extra diligence to find out what God has actually said; but make time to do it. Take a verse, a paragraph, or a short chapter, and think about what you are reading.

Josiah also looked ahead and *"caused all that were present ... to stand to it."* (verse 32) His decision had both a current impact and a future impact on the people and their lives; and God's Word will do the same for you today.

The point is not to adopt one particular system, to read one certain book of the Bible, or to earn some "badge" to piously display to those around you. **Reading God's Word is a matter of informing your thinking with what God has said.** Think about what you read, look ahead to an application in your life, and ask questions so that you know what God has actually said. Reading the Bible takes time—so take time!

◇◇◇

BACKUP CAMERAS

DAY 103

Ezra 3:12 But many of the priests and Levites and chief of the fathers, who were ancient men, that had seen the first house, when the foundation of this house was laid before their eyes, wept with a loud voice; and many shouted aloud for joy:

My Ford Explorer has a fancy screen that does several tasks, like many cars today can do. Whenever you shift into reverse, the screen changes and you can see what is behind you. You may think the backup camera is old news, but it is fascinating to me! I am so fascinated by it that I told my wife that I wish I could see what is behind me while driving down the road.

Evidently, Ford Motor Company considered people like me and created a safety feature that shuts off the camera when the vehicle is shifted into neutral or a forward gear. The only time the backup camera works is when you are backing up. They knew that I would be tempted to watch the screen instead of focusing on the road ahead!

So many people live their lives the same way. They are drawn to past successes or sins, and **they cannot go forward because they are looking behind!** Now, looking back is important, just as using a rear view mirror or backup camera is important. Glancing back occasionally to remember from whence you've come is good; but being held back by your past is not good.

When the captives returned to Jerusalem, they rebuilt the foundation of Solomon's Temple that was destroyed. The Bible says that the older men wept while the younger men cheered. The younger cheered because they didn't know what had been before; the older wept because they were still living in the past and couldn't see the present or future.

Since you cannot relive your past, live in the present by following God's Word today. We are not in the future; we cannot relive the past; we can only engage the present right now. God's perspective exceeds the optimism of the young and the experience of the old. God's Word is not ancient or old-fashioned; It is timeless. And that beats navigating life by a backup camera any day!

◇◇◇

DAY 104

TO QUIT OR NOT TO QUIT?

Ezra 6:14b And they built, and finished it, according to the commandment of the God of Israel, and according to the commandment of Cyrus, and Darius, and Artaxerxes king of Persia.

Is quitting bad?

Are losers always quitters and quitters always losers?

Well, I suppose it depends upon what it is of which you are speaking. For instance, is it not good to quit a bad habit? Is it not good to quit a complaining spirit? Of course it is! Quitting is not necessarily a bad thing to do. Quitters are not necessarily losers. The only way to succeed is to quit things that are bad and start things that are good.

On the other hand, it is wrong to quit something that God has given you to do. It is wrong to quit simply because circumstances become difficult. Anybody can start something; very few are willing to finish something. **We are not fully obedient to God until we fully finish what He has given us to do.**

The Jews who had returned from Babylon to rebuild the temple were discouraged. They were tired. They had encountered great opposition. But in Ezra 6, they finished! The Bible says, *"And this house was finished on the third day of the month Adar…."* What a beautiful statement!

Quitting is a result of our focus on opposition; finishing is a result of our faith in God. Are you ever tempted to quit? If you are doing what God wants you to do, trust the God Who wants you to do it and don't quit!

◇◇◇

REMOTE CONTROL

DAY 105

Ezra 7:6b And the king granted him all his request, according to the hand of the LORD his God upon him.

I like the remote control, don't you? Actually, I like the "control" part better than the "remote." I like to be able to, at a moment's notice, push one button and change things at will. But what happens when someone else (usually another Rice) takes the control? Or the couch swallows the remote so that it cannot be found? When the remote control is out of my control, I feel a little bit nervous!

So it is in life sometimes. We like to have control over our circumstances and our day. We like to "know the score." But what happens when circumstances are out of your hands? **When things are out of your hands, you are still in God's hands.** In the Bible, Ezra is a wonderful example of this truth. He didn't sit on his hands; he didn't wring his hands; he put himself in the care of God's hands! Ezra 7:6 says of Ezra, *"And the king granted him all his request, according to the hand of the LORD his God upon him."* Ezra had everything he needed because he was in God's hands.

You may not have money, but God does—and you are in His hands. You may not have wisdom, but God does—and you are in His hands. You may not have protection, but God does—and you are in His hands. Your single need is not money, wisdom, or protection; your one need is *God*.

Suppose, some junior boys and girls are scared out of their boots when a summer thunderstorm blows across the Ranch. These boys and girls wanted safety, security, and protection. But not one of them yells, "Safety!" What do you think they would cry for? That's right! "Mommy!" Kids instinctively know that everything they need is provided in a person. **When you find things out of control, remember that what you need is completely provided in a Person—God Himself.** When circumstances are out of your hands, you are fine because you are in God's hands. That's the kind of control I want today!

◇◇◇

DAY 106 — GIVE AND TAKE

Ezra 9:12 Now therefore give not your daughters unto their sons, neither take their daughters unto your sons, nor seek their peace or their wealth for ever: that ye may be strong, and eat the good of the land, and leave it for an inheritance to your children for ever.

What are you going to leave your children? Since you don't know *when* you are leaving your children, *what* you leave them is pretty important! And there is much more involved than just leaving money to your children. The people Ezra speaks to in chapter 9 were at risk of leaving nothing valuable to their children. He said, *"Now therefore give not*

… neither take … that ye may … leave it for an inheritance to your children for ever."

What you will leave your children is a predetermined answer. That is, **what you give and what you take determines what you leave.** You do not leave what you want to leave; you leave what you give, what you take, and what you plan to leave. With this truth in mind, **do not give what you cannot afford to lose.** The people in Ezra 9 were giving their daughters to intermarry with the pagans of the land. We can certainly see this in marriages, but the truth extends to our entire lives. Don't give what you can't afford to lose!

Secondly, **do not take what you cannot afford to own.** These people were owning something that God could not own. Again, this started primarily with marriage and *"tak*[ing] *their daughters unto* [their] *sons."* If God can't own it, why in the world would you want to? What kind of blessing can you expect from God for something He would not own? Don't take the world's ways, priorities, or gods—don't take what you can't afford to own!

Thirdly, **leave your kids better than you found them.** I don't just mean how you found them at the hospital! Leave your children an inheritance. Leaving your kids a million dollars but without a lick of common sense and heavenly wisdom is a recipe for failure. Most importantly, leave your kids God. A godly heritage and life of obedience and service to our Heavenly Father is the greatest inheritance you can leave.

What you give and what you take determines what you leave. Don't give what you can't afford to lose. Don't take what you cannot afford to own. Start early; start often. Don't just leave your children money, a house, or other material goods. Leave them the most important and longest-lasting inheritance: God.

DAY 107

ACTIVELY TRUSTING

Nehemiah 4:9 Nevertheless we made our prayer unto our God, and set a watch against them day and night, because of them.

Do you have problems? People problems? Problems at home? Problems that can't wait? Maybe your problem is a ten-year-old problem with blonde hair and blue eyes! What in the world are you supposed to do? Trust God.

What does it mean to totally trust God? Does that mean passivity—sitting back and doing nothing while waiting on God to take care of your problems? I submit to you that totally trusting God means having an "active" rest. I have a friend who is shared the gospel in a hostile environment overseas, and his email address included the phrase "active rest." **Trusting God is a proactive dependence upon Him.** That's a great description of the kind of trust we need to have!

Nehemiah was building a wall that people did not want him to build. In verse 9, he said, *"Nevertheless we made our prayer unto our God, and set a watch against them day and night, because of them."* What did Nehemiah do when faced with opposition and problems? He prayed to God and armed some men to watch the wall! He was *actively trusting* God! He wasn't sitting on his hands; but he wasn't wringing them either. He was trusting God, but he didn't stop what he was doing.

Nehemiah depended upon God, and then he took action, actively trusting in God. Trusting God, in this case, meant setting guards by the wall and arming the workers. **Actively trusting God means that you have a proactive determination to trust God with what you are doing.** The old adage is true: work like it all depends upon you; pray like it all depends upon God. All things being equal, you are the person God is going to use in your workplace and your home. Trust God, take action, and depend upon God for the action you are taking. *That*, my friend, is "active rest."

TO GIVE OR GET?

DAY 108

Nehemiah 5:19 Think upon me, my God, for good, according to all that I have done for this people.

Reading Nehemiah 5-6 is kind of like reading Nehemiah's personal diary! He says in the last verse of chapter 5, *"Think upon me, my God, for good, according to all that I have done for this people."* That sounds pretty bold, doesn't it? He was asking God to notice the good that he had done!

In our world, we are a little more sophisticated about this matter. Instead of going to God about it, we hint to other people and fuss when they don't notice! Was Nehemiah being selfish? Was he complaining to God? Remember this fact about Nehemiah: he was a giver. A person who gives is a person who continually needs. When you give more, you need more. When you need more, you can receive more.

Nehemiah's very honest request to God was based on his dependence upon God. There was a certain amount of desperation when he prayed, *"Now therefore, O God, strengthen my hands."* As adults, we generally do not ask for help if we can keep from it. If you never think you are lost, you will never stop and ask for directions! (I've heard of such things happening…)

Sometimes, like Nehemiah, I remind the Lord that there are certain things I wouldn't need if I wasn't doing what He asked me to do! Some days I need patience that I wouldn't otherwise need if I were not helping His people. Some days I need wisdom that I would not otherwise need if I were not responsible for His Ranch here in Tennessee. I'm sure there are similar needs in your life each day.

Only people who receive help can give it. When you get the help you need from God, you will have the help you need for others. Getting help from God begins with an openness and honesty before God, coupled with humility and dependence. **Follow the example of Nehemiah and ask God for what you need in order to do what He has called you to do.**

DAY 109

WHAT PEOPLE NEED

Nehemiah 8:8 So they read in the book in the law of God distinctly, and gave the sense, and caused them to understand the reading.

Recently, I was reading through a book I had previously read, and there it was. I was absolutely shocked! I could not believe how much of my brilliance was plagiarized in just one chapter! *"I said that! I thought that!"* Well, the reason I was shocked—and shouldn't have been—is that all of that information had been locked up in my head from reading the book ten years ago!

Isn't it amazing what is in our heads? What we read, what we hear, and how we process the information all have a bearing on who we are. Nehemiah 8:8 tells us that Ezra and others rightly divided God's Word (*"distinctly"*) and explained what it meant. God's Word does make sense! **The Bible is not inspired because it makes sense; it makes sense because it is inspired.**

What became of this "preaching service"? The people understood the Bible! People need to hear from God through His Word, not from some contemporary writer, however good he may be. **People need to hear from God.** Notice that these folks gave time and place to the Word of God. They gave a good part of the day to *stand* outside and *listen* to the Bible! Do you give time and place to the Bible? If you are a preacher or Sunday school teacher, do you give time to the Word of God in your sermon or lesson?

Take the Word of God, read it, and explain it to people. **The best way to defend the Bible is to use It!** Tell people what the Bible says! There is a place for being a good communicator and giving attention-grabbing illustrations; but any merit to what you say has to be found from God's Word.

I have found that the more time I spend actually *thinking* about what I've read in the Bible, the more I glean from reading, and the more I have to give others. What people need is what God has said. Give the Bible time, place, and reverence. When God gives you opportunity to teach or preach the Bible, remember that people need to hear from God!

WHO LOVES YOU?

DAY 110

Nehemiah 9:32 Now therefore, our God, the great, the mighty, and the terrible God, who keepest covenant and mercy....

Of all the people in this world, who loves you the most? Why do they love you? If the reason people love you is based on some trait of yours, at some point they will be disappointed. In other words, if people love you because you are attractive, at some point you won't look quite the same as you do now! If people love you because you are smart, you are in a tough spot when you find out there is something you don't know.

In contrast, **God loves you because of who He is, not because of who you are.** That's wonderful! If God loves you because of who you are, at some point everything lovable has an end. How futile to even think that anything I have is lovable to Almighty God! His love for you doesn't change like physical appearance or intelligence can.

Nehemiah 9 reminds us how unlovely Israel was and yet how loving and merciful God was. Verse 32 calls God *"the great, the mighty, and the terrible God, who keepest covenant and mercy."* **God keeps His people because God keeps His promise.** Israel had turned from God over and over again, yet God gave them mercy. His mercy came because of who He is. God is a promise-keeping God. He keeps *"covenant and mercy."*

God is righteous; therefore, He does not lie, say something He does not mean, say something He will not remember, or promise something upon which He will not act. What that means for you and me is this: **since God is a God who keeps His promises, He is a God who keeps *me*.** When I trusted Christ's payment for my sins, it was a done deal. God said, *"Whosoever will may come,"* and I did!

God loves me, not because of who I am, but because of who He is. **Who you love says more about you than it does about the person you love.** How wonderful, then, is the fact that God loves *me*! If you are a child of God, you have a special place in God's heart. Share the good news of God's love with someone who needs it today!

◇◇◇

DAY 111

LIFE'S NOT FAIR!

Esther 3:1 After these things did king Ahasuerus promote Haman the son of Hammedatha the Agagite, and advanced him, and set his seat above all the princes that were with him.

Life isn't fair. Have you noticed this fact? Actually, I'm glad it is not! I am glad that God is in control, not the events of life or happenstance. Though you will not find the name of God written in the book of Esther, God is imprinted on every page, every chapter, and every verse. You can't find a more obvious example of God's providence than the story of Esther!

Sometimes life seems like it is governed by impulsive, undiscerning authorities. Such was the case in Esther's day. King Ahasuerus loved Esther and gave her his kind favor, yet he also was kind to the wicked Haman who desired to slaughter an entire race of people! As you may know, the position of queen wasn't exactly like winning the lottery; things didn't end so well for the previous queen! It would have been easy for Esther to see life as cruel and unfair and governed by impulsive authorities. Have you ever felt the same way about *your* life?

Sometimes life seems like it is governed by random events. Things just do not seem to have rhyme or reason. The book of Esther is full of what seems like random events. Esther "just so happens" to be chosen by the king, Mordecai "just so happens" to save the king's life, and Haman "just so happens" to be near the king's court when Mordecai is rewarded. Can't you see how

Esther could have viewed her life as a series of random events? Does your life ever seem this way?

You do not find God's name *one* time in the story of Esther. You may not see His name, but you *can* see His face on every page! What seemed like the random events and selfish wishes of an impulsive leader were all tools in the hand of God. Life isn't fair. It wasn't in Esther's day, and it won't be today. Even when you cannot see God in the events of your life, you can know that He is in control. **When life is out of your control, you can know that God is in control of your life.** While you do not know the end of the story of your life, trust the God that made this world and made you!

TURNED TO THE CONTRARY

DAY 112

Esther 3:7 In the first month, that is, the month Nisan, in the twelfth year of King Ahasuerus, they cast Pur, that is, the lot, before Haman from day to day, and from month to month, to the twelfth month, that is, the month Adar.

In Esther, if you see God's face on every page, you do not have to see His name. Even though God's name does not appear in the entire Book of Esther, you can't help but see His face and see His hand at work. Some folks are very tripped up by the fact that God's name does not appear at all; but if I see a photo of my grandfather, I know it is him, whether or not I see his name on paper. In the same way, I can know God was in control of the events in the book of Esther, even though I do not see His name written.

The momentum—the turning point—in the story came when Mordecai's deeds in saving the king's life were made known. As the Bible demonstrates, events were *"turned to the contrary."* (Esther 9:1) The gallows were turned on Haman, and Mordecai was promoted in the king's court.

The story of Esther, Mordecai, and Haman is more thrilling than fiction, yet it really happened. The series of events seems like "dumb luck." Does life ever feel that way to you? Have you recently thought, "Why today?" or, "Why did *this* happen?"

The word *Pur* means "a lot; one dice." *Purim* is the plural—"two or more dice." You can read about the Feast of Purim in Esther 9, but the Jews were celebrating God's providence. The name literally means "Feast of Chance." Jewish people still celebrate Purim in the month of March.

So what does all this mean for us? Proverbs 16:33 says, *"The lot is cast into the lap; but the whole disposing thereof is of the LORD."* In other words, **you don't have to understand everything about today; God is in control.** What sometimes may seem like *sheer, dumb*—and sometimes *bad*—luck does not mean that God is not in control. He is present, and you can trust Him with what you don't know.

◇◇◇

DAY 113
GOD'S TIMING

Esther 6:4 And the king said, Who is in the court? Now Haman was come into the outward court of the king's house, to speak unto the king to hang Mordecai on the gallows that he had prepared for him.

When my daughter was ten years old, she asked me, "Daddy, why does the Bible say life is so short, when it seems so long?" Ah, the perspective of a ten-year-old! Life seems so long when you are ten, but it seems short when you are eighty. As humans, our perspective on time can vary widely; just imagine how much our perspective differs from God's! II Peter 3:8 says, *"But, beloved, be not ignorant of this one thing, that one day is with the Lord as a thousand years, and a thousand years as one day."*

We are consumed with time—long, short, too much, not enough, etc.—but **God is the Master of *timing*.** Timing is important, and you can see that over and again in the book of Esther. At the very same time Haman came to the king for permission to hang Mordecai, the king was looking for someone to give advice on honoring Mordecai! Because God is in control, He is in control of timing.

Don't get so caught up in time today that you forget about timing. This is not an excuse to be late or lazy, but don't chase away your day. **God is not hurried by time, and He is not worried.** Yet how many times are we hurried and worried about time? God is the Master of timing. Whether you wish something would last longer, or you wish something would end already, remember that God is in control and is concerned about timing. If you can trust God with your life, you can trust Him with the timing in your life!

LEAVING

DAY 114

Esther 10:3 For Mordecai the Jew was next unto king Ahasuerus, and great among the Jews, and accepted of the multitude of his brethren, seeking the wealth of his people, and speaking peace to all his seed.

Do you love to leave? Maybe it comes from the many years of traveling, both as a child and now with my own family, but I love to leave! The truth is, we are all living lives in which we are continually leaving one place or another. The last glimpse we get from the book of Esther about this man Mordecai is a fascinating epitaph: *"seeking the wealth of his people, and speaking peace to all his seed."* We learn that **the way you live determines the way you leave.**

There will come a time when you make your departure from this earth. The question is not *if* you will leave; rather, the question is, what will you leave and how will you leave it? **What you leave and how you leave it is determined by how you live.** Mordecai left exactly the way he lived! He was doing the exact same thing in the palace that he was doing under persecution a few chapters before! *"Seeking the wealth of his people."* He was a man of courage, conviction, and integrity; and that is precisely how he left!

If you want to leave the right thing behind, you must live the right way. Living the right way begins now, specifically today and in the list of things you have on your to-do list right now. None of us knows when he is going to leave, but we can make the most of the time we have right now. What do you want to leave behind? Well, how you live today will answer the question of *what* you leave *when* you leave. **Take advantage of the time you've been given in order to prepare for when you leave!**

◇◇◇

DAY 115

OUR DAYSMAN

Job 9:32-33 For he is not a man, as I am, that I should answer him, and we should come together in judgment. Neither is there any daysman betwixt us, that might lay his hand upon us both.

Has God blessed you? How has God blessed you? If you only base your answer on the balance in your checking account or the number of things you possess, you might have a problem! Consider Job. He had lost his money, his family, his house, and his cattle. In our minds (and in the minds of Job's friends), if we have things, God must love us; and if we have nothing, God must dislike us.

Job couldn't see God, understand God, or even answer any of God's questions. What he needed was a "daysman," a go-between. **Thank God for *the* Daysman, the Lord Jesus!** Between a holy God and sinful man is the God-Man, Jesus. That is Who Job needed to find, and that is Who we need to find.

I Timothy 2:5 says, *"For there is one God, and one mediator between God and man, the man Christ Jesus."* God sent His Son to live in human form. We have a daysman to "go between" for us. **Instead of looking for a reason for our lives' events, we ought look for this Person Who has the world in His hands!** You may not know every reason or every cause for what is happening in your life today, but you can know the God behind every situation. Look for Him today instead of looking for a reason.

LISTEN. LOOK. GIVE.

DAY 116

Job 12:1-2 And Job answered and said, No doubt but ye are the people, and wisdom shall die with you.

Let's imagine that we are doing some hiking and mountain climbing out West. As we approach the massive elevation change before us (a nearly vertical pitch), I offer to take the lead and climb first. Before climbing much further, I reach down almost directly below me to pull you up. As we lock hands, I place my feet on top of your shoulders, so that I am pulling you with my hands yet pushing with my feet. Instead of a firm foundation on *terra firma*, I am using your shoulders to stabilize myself!

In our little imaginary story this morning, how much progress are we likely to make? Not much! Our story illustrates the truth that **you cannot put someone down and pull someone up at the same time.** It's impossible to do, whether you are mountain climbing at Yosemite or dealing with people. Job had friends who tried to pull him up while they were putting him down.

How can we pull people up and not put them down? First of all, **listen**. Sometimes wisdom is best expressed without a word. The best thing to say may be nothing at all! Second of all, **look**. Look at yourself! Job said, *"But I have understanding as well as you; I am not inferior to you...."* Don't look down on people who need help, not realizing that there are areas in which *you* need help also. Third of all, **give**. Give them the truth of God's Word. Give the truth from a heart that wants to help in a meek and humble spirit. Ultimately, I am not helping anyone if I don't give them the truth!

How do you help people who are struggling? Well, you certainly don't help them up by putting them down! When helping people, we need a helpful spirit and a willingness to listen, look, and give. **Complement the truth of God's Word by listening to others and looking at yourself.** We may not go mountain climbing today, but I hope we can help people who need to be pulled up today!

◇◇◇

DAY 117 — GOD UNDERSTANDS YOU

Job 23:10 But he knoweth the way that I take: when he hath tried me, I shall come forth as gold.

There is probably not a man reading this that has not said, "I just do not understand women!" Likewise, there is probably not a woman reading this that has not said, "I just do not understand men!" Well, I have good news. You both are correct!

Not understanding people can be very frustrating at times! Some people you just can't figure out, no matter how hard you try. **The only thing more frustrating than not understanding people is not being understood by people.** That is hard! Sometimes it can feel like nobody "gets it." Sometimes it can feel like nobody understands you.

Friends did not understand Job, and Job did not understand God. He said, *"I go forward, but he is not there; and backward, but I cannot perceive him: On the left hand... I cannot behold him: he hideth himself on the right hand, and I cannot see him."* Have you ever felt that way? You look left and right, up and down, and you cannot seem to find God.

I do not need to understand God's ways to be assured of the fact that He knows mine. Even when I cannot "find" God, He knows the way I take. I don't always understand God, but He always understands me! God is righteous, all-knowing, and all-powerful, so trust Him today. Even when no one understands you, the One who made you does!

THE POWER OF INFLUENCE

DAY 118

Job 26:4 To whom hast thou uttered words? and whose spirit came from thee?

If you want to know a family's innermost secrets, just offer to babysit the kids! It is interesting to see the influences that we pick up, and there is no better way to manifest those influences than through our kids. Children and adults alike are always speaking something to someone from somebody else. Influence is important!

We have a tendency to explain away our responses to people and situations by blaming our region of the country, our family heritage, or life's circumstances. **When you are "squeezed" by life, the real you is going to come out!** The real you is made up of the influences you've allowed into your life.

It ought to be that when you are squeezed, the Lord comes out. I want to be so filled with Him that when the pressure is greatest, the Lord Jesus is revealed. The question of Job is a good one: *"Whose spirit came from thee?"* Whose influence will be revealed in you today?

Don't be defined by the worst of your region, your family, and your circumstances. Be a vessel for use by a God Who supersedes all of that! God is greater than the worst secrets and habits of any family. Just don't ask to babysit my kids anytime soon!

ONE THING YOU CAN'T LOSE

DAY 119

Job 27:5 God forbid that I should justify you: till I die I will not remove mine integrity from me.

What do you have that cannot be lost or taken away from you? Well, anything you could have, Job did have; and anything you could lose, Job lost! He lost health, wealth, family, and even his reputation with his friends. In all of his trouble, Job maintained integrity and consistency. He said, *"God forbid that I should justify you: till I die I will not remove mine integrity from me."*

While you can lose all the things Job lost, you can keep the one thing Job kept: integrity. **No one can take away your integrity!** Satan tried to attack Job's integrity. Job's wife questioned his integrity by saying, *"Dost thou still retain thine integrity? curse God, and die."* Job's own friends accused him of compromising his integrity. Despite all of this, Job could say, *"Till I die I will not remove mine integrity from me."*

There should only be one "you." There should not be a "church you", a "free time you," and a "private you." In essence, integrity is when there is only one "you"! Integrity means a singleness of heart and action, not a duplicity.

Integrity can be sold, but it cannot be bought. You cannot buy a good reputation, but you sure can sell it. When you sell your integrity and your good name, you sell it for a price that you cannot afford!

Integrity is absolutely vital and basic to everyday life. **Nothing is worth anything without integrity.** What good is money if the guy who owns it is not honest? Would you trust him to invest *your* hard-earned money? What good is power without integrity? Have you ever heard of a corrupt politician or government leaders? Proverbs 22:1 reminds us that *"a good name is rather to be chosen than great riches."*

Ultimately, someone can take everything you have, except for your integrity. There should be only one "you." Don't sell your integrity for anything!

THE AGONY OF A GUILTY CONSCIENCE

DAY 120

Job 31:6 Let me be weighed in an even balance, that God may know mine integrity.

Think about the worst pain you have felt, apart from physical pain. For some, anxiety is the worst. For others, longing is. Perhaps it is the pain of losing something or someone. Whatever pain you might experience in the next year, there is one kind of pain you do not have to experience: the pain of a guilty conscience.

Of all the things you suffer in life, a guilty conscience does not need to be one. While it is true that you will never be able to pull one over on God, the devil doesn't have to be able to pin anything on you! God will never owe you, but you need not ever owe the devil.

While you may not have control over the circumstances you will face, you can have the peace of a clear conscience. There is no agony worse than a guilty conscience! On the other hand, there is no foundation for peace greater than just being clean before God.

The greatest agony? A guilty conscience. The greatest peace? A clean conscience. God's power is absolute; His holiness is absolute; and if you have a clean slate before Him, you will be just fine. Job was convinced of this truth. Are you?

WHO OWNS YOU?

DAY 121

Job 42:2 I know that thou canst do every thing, and that no thought can be withholden from thee.

How many problems do you own? Your number of problems probably correlates to your number of possessions! The more "stuff" you have, the more likely you are to have problems. However, if God owns you, He owns your problems.

Job was reminded of this truth when God called him "my servant Job" four times in the last chapter of Job. **Because God owned Job's possessions, He owned Job's problems.** In short, God owned Job.

Oftentimes, we hold what we own with a tight grasp. If you own your possessions, guess what? You own your problems! When you remember God's ownership of you, wonderful balance and stability come to your life. God is in control.

God reminded Job—and He reminds us through Job—that in a moment, everything a man possesses can be taken away. God owns it all to begin with! God's ownership gives perspective to our problems. God owned Job and Job's problems. That is not a bad thing at all!

It might be a good idea to begin each day by reminding yourself that God owns you. That means that He owns everything you have, and He owns your problems. Don't struggle through each day and each trial trying to own your problems. **In prosperity and in poverty, God owns it all. God owns *you*.** Job learned that, and we ought to as well!

◇◇◇

DAY 122
LEND ME YOUR EAR

Psalm 5:1-3 Give ear to my words, O LORD, consider my meditation. Hearken unto the voice of my cry, my King, and my God: for unto thee will I pray. My voice shalt thou hear in the morning, O LORD; in the morning will I direct my prayer unto thee, and will look up.

The most important truth about prayer is that God hears. Notice the words David uses in the first three verses of Psalm 5: *give ear, hearken, my voice, hear*. C.H. Spurgeon said that the object of prayer is God's ears. That is true, and that is what we see in David's prayer in Psalm 5.

It is wonderful that God not only knows the words we pray but also knows our thoughts. David said, *"Consider my meditations."*

Sometimes people hear your words (as in a public prayer). But whether or not people hear you, the most important thing about prayer is God's ear.

Notice also in verse 3 that David says, *"I will direct my prayer unto thee, and will look up."* You can't look up if you are on top of the pile. When you come to God and look up, that is the way it should be. A man that hasn't figured out that he is needy will not look up. But the first thing you should do when you get out of bed is "look up."

Are confidence and humility opposites in your mind? Well, if by confidence we are talking about how great you are, then humility is nowhere to be found. That type of confidence is just begging to be crushed. But truthfully, confidence and humility are not mutually exclusive. True confidence comes from true humility, and the best way to demonstrate humility is to "look up" to the God of heaven.

◇◇◇

A WORD ON WORDS

DAY 123

Psalm 12:6-7 The words of the LORD are pure words: as silver tried in a furnace of earth, purified seven times. Thou shalt keep them, O LORD, thou shalt preserve them from this generation for ever.

Talk is cheap. Actually, talk is cheap when it is fickle. If I tell you one thing today yet do something entirely different tomorrow, the words I said are mighty worthless! In our world today, it is easy to feel just like David in Psalm 12:1—*"the godly man ceaseth; for the faithful fail from among the children of men."* Sometimes after hearing all the chatter on the radio, television, and online, I wonder what happened to all the people who make sense and speak truthfully!

During election time, I received several mailers from various candidates. It was the most amazing thing. Every single candidate was a great person and the best candidate for the job! The psalmist David must have had similar thoughts about the talk he was hearing. *Vanity, flattering lips,* and *double heart* (verse 2) are not commendable descriptions!

In contrast to the empty double-talk of people, *"the words of the LORD are pure words."* **His words are always pure, always right, always perfect.** That is why His Word (the Bible) is so precious and so vital to our lives. This morning, you can read the very words of God and know that they are from God Himself!

Don't you appreciate a man who keeps his word? In the midst of the noise of everyone's opinions, those who keep their word and speak the truth rise to the top in my mind. Do you know whose words are *always* right and true? God's words are, and He *always* keeps His word!

So today, as the kids' song so aptly notes, "Be careful, little tongue, what you say." Don't let your speech be cheap and empty today. In addition, thank God that you can read the Bible with confidence, knowing that His words are pure and preserved. **God is reliable and trustworthy, and what He has said *will* come to pass.** Don't just thank Him for that truth; get to work reading His Word!

◇◇◇

DAY 124

THE MORE THINGS CHANGE...

Psalm 16:8 I have set the LORD always before me: because he is at my right hand, I shall not be moved.

If you asked the average person, "Are you happy?" what answer would you expect? Regardless of the answer, most people would base their answer on what they have, where they are, or what they are doing. The truth is, all of those things are subject to change. Circumstances change all the time. Our feelings change. Our thoughts change. But Psalm 16 reminds us that God is immovable!

The more our world changes, the more **we need the stability that only comes through God.** The psalmist reminds us that living in God's presence is a decision you must make. *"I have set the LORD always before me … in thy presence is fullness of joy …."*

Whatever you are "chasing" today, remember that peace is not found in what you have, where you are, or what happens to you. Even if you attained everything you wanted and life was smooth sailing, yet the Lord was distant, you would not find the comfort you are seeking. **Peace and joy are found *in God's presence.***

It is not wrong to be fulfilled and satisfied; it is folly to seek those things apart from a Person. Don't look to a place or a possession when what you need is found in a Person!

◇◇◇

PAYBACK!

DAY 125

Psalm 18:24 Therefore hath the LORD recompensed me according to my righteousness, according to the cleanness of my hands in his eyesight.

Our God is a rewarder, and you can count on Him to recompense (pay back) both you and those around you. What if a person seems to be getting away with doing something wicked? God will repay. What if someone is doing right but not receiving the credit for it? God will repay. **He is always rewarding!**

What kind of reward are you receiving from God right now? Is He merciful, or is He opposed to what you are doing? You will be repaid; whether that is a good or bad thing depends on what you are doing! A "payback" is a response to something that has already been paid out. The Bible promises in verse 26, *"With the pure thou wilt shew thyself pure; and with the froward thou wilt shew thyself froward."*

Verse 30 reminds us, *"As for God, his way is perfect…."* Rest in the truth that whether God is rewarding or leading, He will always be right. **His way is always perfect, even though it does not always appear to be.** We can be so focused on what is directly in front of us that we forget that God can see the entire way clearly. You can have confidence in God's way because it has been tested (*"tried"*). Your way in life can be perfect too if you will follow God's way.

This morning, know that God is everything you need and everything you are not. He is a rewarder, and His way is always perfect. Whether or not today is a "good" day by your own standards does not change the fact that you have a good God who will always do right and whose ways are always right. That makes every day a good day for a child of God who is following God's way!

◇◇◇

DAY 126 — INSUFFICIENT POWER

Psalm 20:7 Some trust in chariots, and some in horses: but we will remember the name of the LORD our God.

Whatever you need, God is able to provide it. He is in control, and we do well when we remember that truth and trust Him. **God will do what we cannot do when we do what we should.**

Chariots were the most powerful weapon known in the psalmist's day. That is why the Syrians, by way of example, attributed their dominating success to the flat land (where the chariots were advantageous). I Kings 20 reminds us that God is the God of both the hills and the valleys. The victory you need is not found in a thing or a place; it is found in a Person.

Do you know anyone that is currently relying on horses and chariots for victory? Does your neighbor have a chariot in his garage? That is the point! No one fears a chariot these days. What the Syrians trusted in has now become just a relic with no real power. How futile! In spite of our advancements in technology, military power, and transportation, we still forget the Source of our victory too!

We are sitting on the greatest sum of human knowledge any generation has every possessed. Things we take for granted today were major ordeals a hundred years ago. **What has not changed is our need to trust (*"remember"*) the Lord.** If you trust in the most cutting edge technology today, you are trusting in something that will soon be "old-fashioned." God is not old-fashioned or antiquated; He is real-time right now. His power is never diminished.

Don't place your trust in something today that will not last, will not keep up with the times, and will not succeed. Trust the LORD God Who spoke the world into existence, Who made you, and Who cares for you. He will be just as powerful tomorrow as He is today, and He is just as powerful today as He ever was.

HURRY UP AND WAIT!

DAY 127

Psalm 25:5 Lead me in thy truth, and teach me: for thou art the God of my salvation; on thee do I wait all the day.

Sometimes waiting can be very uncomfortable. For the most part, we are comforted by noise and speed. It does not matter where we are going or what is being accomplished, as long as noise is made and things happen! When it comes to driving, I would rather make good time going nowhere than to stop and ask directions to get where I am supposed to be.

Waiting is the exact opposite of this sentiment. Regardless of your temperament, waiting does not come naturally. Psalm 25 reminds us just how important it is to wait. What does it mean to "wait on God"? It means *to trust* God! Psalm 25:2 says, *"O my God, I trust in thee."* Trusting God is one thing; patiently trusting God is another.

Trusting God is relatively easy when you think you know how He is going to provide. It is a horse of a different color to trust God when you can't see how God will possibly answer or provide! Are you trusting God? Better yet, how are you trusting God?

You will never be late when you decide to wait. **When you choose to wait on God, you will always be in the right place, at the**

right time, and ready for the right thing. God is worthy of our waiting. He is "in it for the long haul," and He knows what is best. So, hurry up and wait on God today!

DAY 128 — STEPPING AND STANDING

Psalm 26:11-12 But as for me, I will walk in mine integrity: redeem me, and be merciful unto me. My foot standeth in an even place: in the congregations will I bless the LORD.

Imagine getting on a bus or plane, only to find out that you got on the wrong one! While two planes from the same airline may look similar, they could be heading for opposite ends of the country (or beyond)! Two buses may both say "Greyhound" on the side; but Amarillo, Texas, is completely different than Manhattan, New York. You'll not end up at the right place if you do not have the right path.

In the same way, purity before God gives us confidence for life. When I have a clear conscience before God, I have a confidence for life that can come no other way. This confidence is not a matter of confidence in my ability; it is a matter of my obedience to God!

The psalmist said in Psalm 26, *"I will walk ... my foot standeth...."* He was confident about his future because of where he was standing that day. **Where you step today determines where you stand tomorrow.** If you step on the right road, you will end up in the right place! If you step onto the right airplane, you will arrive at the right destination.

You may not be where you want to be spiritually, but if you will take a step in the right direction, you will be one step closer than you were! Your kids may not be where they ought to be; but if you will help them step on the path of obedience, they will end up in the right place later. **Roads determine destinations.**

You may be miles from home, and you may not be able to see it from here; but if you will step on the right road, you can be confident that you will end up at the right place. Choose the right path today so you will arrive at the right destination later!

"Judge me, O LORD; for I have walked in mine integrity: I have trusted also in the LORD; therefore I shall not slide."

FLEEING TO OR FROM?

DAY 129

Psalm 55:22 Cast thy burden upon the LORD, and he shall sustain thee: he shall never suffer the righteous to be moved.

Have you ever wanted to just fly away from the trouble you are facing? **The natural instinct of most creatures, when faced with hostility, is to flee or fight.** That is true of animals you might cross paths with today; and that is true of us when faced with hardship. Our first tendency is to flee or to fight. However, in Psalm 55 David reminds us that instead of fleeing (moving), we can be settled.

David himself felt like fleeing: *"And I said, Oh that I had wings like a dove! for then would I fly away, and be at rest."* Sometimes I think that birds have it made! Birds don't pay bills, have flat tires, or face nearly the difficulties I can face every day. If they don't like the weather, they can just fly south! For Christians, there are times when it is appropriate to flee or to fight, but if you live your whole life simply looking to flee or to fight, you really are missing out on God's desire for His children.

Instead of fleeing or fighting, David said to *"cast thy burden upon the LORD, and he shall sustain thee…."* **The natural instinct for a child of God should be to ask for help!** You can try to flee from trouble when life is pressing down on you, but you will spend your time worrying and scheming. You can't sustain that type of living! In contrast, when you cast your burden on the Lord, the Bible promises, *"He shall sustain thee: he shall never* *suffer the righteous to be moved."* **Are you bearing your burdens alone or are you casting them on God?**

God can be your shelter when your tendency is to hide. Don't live your life fleeing from the trouble that life sends your way. *"Cast thy burden upon the LORD, and he shall sustain thee: he shall never suffer the righteous to be moved."* Don't move away from your problems; move your problems to the One who can take care of them!

◇◇◇

DAY 130

ONLY GOD

Psalm 62:5-6 My soul, wait thou only upon God; for my expectation is from him. He only is my rock and my salvation: he is my defence; I shall not be moved.

When SEAL Team Six stormed the bin Laden compound in Pakistan, they had both a plan and a contingency plan. They had a backup plan for almost every conceivable problem they might encounter. In fact, one of the strategists confirmed that the SEALs routinely practice exactly these contingency plans. They plan on having problems, and they plan how they will adapt and overcome.

While contingency plans are fine for the Navy SEALs, they are not fine for you! If you enter today with a backup plan in case God does not provide, you are already defeated. The truth is, **the best place to be is where God is your *only* option.**

It can be an unsettling time when God is your only option; but according to Psalm 62, that is the perfect situation. C.H. Spurgeon called Psalm 62 the "Only Psalm." You can't escape the word *only* and the truth that God is the only God! Is God your only option?

A change of plans, a lack of power, or a lack of friends may conspire against you; but a silver lining is that God becomes your only option. You cannot always predict what life will throw at you. But you can know that He is your rock and your defense, and you can say with the psalmist, *"I shall not be moved."*

Today, let God be your only option for what you will face. No "plan B." No "plan C." Only God. You may not have all the answers, but when God is your only option, you have the One who does have all of the answers! He really is all you need!

◇◇◇

LOOKING BACK

DAY 131

Psalm 78:9 The children of Ephraim, being armed, and carrying bows, turned back in the day of battle.

Each year we remember the terrorist attacks on the United States on September 11, 2001. There has been some discussion in recent days about whether or n ot we should remember that day's events anymore. Should we "put it to rest," as some have suggested? Should we forget 9/11 in favor of a domestic agenda that neglects the need to protect our land?

Did you know that we are not the only people who have faced such a decision? Dwelling on the past never helped anybody, but moving forward effectively requires a remembrance of where you have been. In other words, you can't know where you're going if you don't know where you've been! Psalm 78 reminds us that **those who never look back are bound to turn back.**

The children of Israel repeatedly forgot what God had done. Verse 11 says, *"And forgat his works, and his wonders that he had shewed them."* Verse 42 tells us that *"they remembered not his hand, nor the day when he delivered them from the enemy."* Psalm 78 is full of forgetting! What came of their forgetting? They *"turned back in the day of battle"* and *"turned back, and dealt unfaithfully like their fathers…."* (verses 9 & 57)

It is amazing how we, as a nation and as individuals, keep repeating the same mistakes. Human nature hasn't changed, and the answer to man's problems hasn't changed! The warning from

God's Word this morning is clear: **if you never look back and remember what God has done, you are destined to turn back.**

Even the children of Israel, who saw the miracles of the Old Testament firsthand, turned back when they forgot what God had done in their lives. Today, don't forget what God has done and where He is guiding you. When you fail to look back, you are bound to turn back. Don't do either—look back!

◇◇◇

DAY 132

WE CALL, GOD ANSWERS

Psalm 81:7 Thou calledst in trouble, and I delivered thee; I answered thee in the secret place of thunder: I proved thee at the waters of Meribah. Selah.

I think it's true that each of us needs a person who we can call and we know will answer. It may sound insignificant, but **nothing is more valuable in time of need than someone who will answer when you call!**

God Almighty is speaking in Psalm 81, and He says, *"Thou calledst in trouble, and I delivered thee; I answered thee...."* I love the fact that **when we call, God answers!** When do we normally call? I wish I could say otherwise, but we call when we are in trouble. The truth is, we want deliverance; God wants our attention.

Our problem is not our problem; our problem is not calling out to God about our problem. **Look up when you are in trouble!** As this verse points out, sometimes God answers us by proving us. He may deliver or He may test, but both are in response to your calling. Sometimes God tests you by withholding an answer; sometimes God tests you by giving an answer. Whether He delivers or tests you, **He always does what is right and what is best.** Thank God that He always answers!

God help us to respond correctly when we find ourselves in trouble. Remember to look to God and call on Him. Whether

He answers by delivering or by testing, know that He graciously and mercifully answers when we call!

WHO ARE YOU SERVING?

DAY 133

Psalm 84:10 For a day in thy courts is better than a thousand. I had rather be a doorkeeper in the house of my God, than to dwell in the tents of wickedness.

I have seen some amazing homes in my years of traveling. I've driven by the Hearst Castle in California, beautiful homes around Rodeo Drive in Beverly Hills, and expansive homes in the mountains on the East Coast. What is the first thing that comes to mind when you see an amazing home? I wonder who in the world lives there! You also probably wonder what they do in order to afford such a house!

What you do in life is certainly of consequence. But the psalmist reminds us in Psalm 84 that **who you serve is more important than the position you hold.** That is, serving God is more important than the capacity in which you serve. Whether you type on a computer, cut wood with a saw, wrangle horses and cattle, or serve some other way, Who you serve is what is most important!

Don't lose sight of Who you serve while you are serving. We can get so wrapped up in God's will for our lives that we lose sight of the One we are serving. Don't get so stuck on what you are doing that you forget Who you are serving!

If you will serve the right Master, you will reap the right reward. The Bible promises, *"No good thing will he withhold from them that walk uprightly."* Many times we are concerned about what we do and what we get from doing it. Instead, focus on serving the right Master and trusting Him for the right reward for your efforts.

What you specifically do for the Lord *is* important, but it is not the most important thing. *Who* you are serving is the most important thing! I know a guy who changed light bulbs for a living. That doesn't sound impressive, at least until I tell you that he changed light bulbs in the White House! Who he is serving (the President) elevates the importance of what he is doing (changing light bulbs). More than any president, king, or ruler, you serve the Living God, and you can trust Him with the reward. Remember Who you serve today!

◇◇◇

DAY 134 — GOD IS JUDGE

Psalm 94:1-2 O LORD God, to whom vengeance belongeth; O God, to whom vengeance belongeth, shew thyself. Lift up thyself, thou judge of the earth: render a reward to the proud.

If you are guilty of robbing God of anything, what would it be? That question is a bit repulsive, isn't it? But robbing God is not an impossible concept. Psalm 94 reminds us that **robbing God of enemies that belong to Him is never a wise thing to do.**

It is easy to think, "If God would take care of this, I wouldn't be tempted to do it myself." Well, that is exactly your problem! God's timing is impeccable. He is not early; He is not late; He is right on time. **You can trust God with your enemies and problems, just as surely as you can trust Him with your money, your life, and your eternity.** This, too, is a matter of faith.

God is called the *"judge of the earth"* in verse 2. That means, first of all, **He will reward the proud.** Don't steal God's judgment of the proud; He will take care of them! Secondly, **God will take care of His own.** Verse 12 says, *"Blessed is the man whom thou chastenest, O LORD, and teachest him out of thy law; That thou mayest give him rest from the days of adversity, until the pit be digged for the wicked."* There is a difference between God's judgment of the wicked and His chastisement of His children. On a human level, there is a difference between shooting an intruder and spanking a son. God takes care of His children!

Thirdly, as Judge of the earth, **God defends the innocent.** Verse 22 says, *"But the LORD is my defence; and my God is the rock of my refuge."* When you are innocent, and you do not try to steal God's vengeance, God will be your Defender. **God is Judge, and it is a wonderful thing when we recognize this and allow Him to be as much in our lives!**

◇◇◇

OUR GREAT GOD

DAY 135

Psalm 95:1 O come, let us sing unto the LORD: let us make a joyful noise to the rock of our salvation.

It is very hard to magnify God when you are the biggest thing in your life. It is equally as hard to magnify the Lord when the problems you face are the biggest things in your life. Psalm 95 reminds us that we have an obligation to praise and thank the Lord.

We are to *"sing unto the LORD"* and *"make a joyful noise."* If you are like me, the second one is easier than the first! We are to praise the Lord and *"come before his presence with thanksgiving."* **When you are magnifying God, self will diminish along with the problems that conspire for your attention.** The larger God becomes, the smaller I and my problems become! That may be easier said than done, but it *is* the reality God wants us to see in our lives.

So why are we to praise God? Because **He is a *great* God** (verse 3). If God is a great God, that means your problems are not bigger than He is. Conversely, if you spend your life trying to make yourself great and jockeying for stature and standing, you have nothing to help when life's problems are greater than your ability.

Not only are we to praise God because He is a great God, but we are also to praise Him because **He is a *"great King* above all *gods."*** If God were not above all gods, He would cease to be God! God isn't just a good option or the best option; He is the only option.

God is a great God; God is a great King above all gods; and thank the Lord that **He is our God!** Verse 7 says, *"For he is our God; and we are the people of his pasture, and the sheep of his hand…."* Sheep are defenseless and need direction. A sheep that thinks he can run his own pasture is destined for trouble. We are the *"sheep of* [God's] *hands."* Thank the Lord for that truth; rest in it.

Magnifying the Lord is difficult when you are the biggest thing going in your life. This morning, why not take time to praise God for Who He is and thank Him for what He has done. He is *"a great God, and a great King above all gods."* Praise the Lord that He also is *"our God"*!

DAY 136

SERVE THE LORD WITH GLADNESS.

Psalm 100:2 Serve the LORD with gladness: come before his presence with singing.

Is there a difference between *working* and *serving*? Sure! They are both basically the same, but work is another word for labor. You can *work* for food, for friends, or for funds, but work is just…work! **Serving** is a specific kind of work: work that indicates a direction. You can work without regard for others, but you cannot serve that way. Serving is labor for somebody else. It will make all the difference in the world today if you are *serving* instead of just *working!*

There is great profit for you and for others if your labor today is serving. But there is a difference between just *serving* and **serving the Lord.** You can serve a person, a ministry, or a cause, but serving the Lord is something altogether different! Serving will affect the way you work, and serving the Lord will affect how you serve.

Ephesians 6:6 says servants are to obey their masters *"not with eyeservice, as menpleasers; but as the servants of Christ, doing the will of God from the heart."* If you are simply serving, the temptation will be to serve only when your boss is watching or will find out. Serving the Lord is another story! Serving the Lord means *"doing the will of God from the heart."*

Is there a difference between *serving the Lord* and **serving the Lord with gladness**? You'd better believe it! Have you ever met someone who was serving the Lord but wasn't glad about doing it? *How* you serve is important, just like *Who* you serve is important. God tells us here to *"serve the LORD with gladness."*

In just five short words, the Bible gives us a game plan for today. When you put all three parts together (serve–the Lord–with gladness), you will find a day with purpose, delight, and the blessings of God, no matter what today may bring.

"Serve the Lord with gladness."

THE LIFE-GIVING WORD

DAY 137

Psalm 119:159 Consider how I love thy precepts: quicken me, O LORD, according to thy lovingkindness.

Psalm 119 is all about God's Word! You find references to *statutes, commandments, precepts, laws,* and *judgments* throughout the entire psalm and in almost every verse. The Word of God is a number of things to us and a number of things for us. This psalm reminds us that God's Word teaches us, guides us, instructs us, and protects us. It is vitally important to our daily lives!

One word that stood out to me during a recent reading of Psalm 119 is the word *quicken*. *Quicken* means "to make alive; to revive." Verse 25 says, *"My soul cleaveth unto the dust: quicken thou me according to thy word."* Verse 37 says, *"Turn away mine eyes from beholding vanity; and quicken thou me in the way."* Verse 50 declares, *"This is my comfort in my affliction: for thy word hath quickened me."* Those are just three references; there are many more like them throughout the psalm!

The bottom line is, if you are not in the Word of God, you are not going to have "life" in your Christian life. **There is no life outside of God's Word.** There are many ways to get God's Word each day in a meaningful way. I've recently found the Bible recorded on audio to be a help to me. Having the Bible on audio has helped me "get" what the Bible is saying, even when I have to rewind it several times! I have enjoyed hearing the Bible; but other times, I can see Bible truths by reading what I miss by listening. The point is this: **find a time and way to get God's Word into your life.**

The Bible is a supernatural book, not a magical book. It will not change your life by happenstance or default. The Bible makes sense; and when you know what God has said and obey it, your life will have life. God's Word is vital to your daily existence as a child of God, so much so that God devoted a whole psalm in the Bible to the Bible!

DAY 138 — WAITING EYES

Psalm 123:1 Unto thee lift I up mine eyes, O thou that dwellest in the heavens.

You can tell so much by who it is that gets the eyes in any given situation. For instance, if you are in a group of people, many times one person will command the attention of the whole group when he speaks. Everyone in the group defers to the person speaking; and in a sense, everyone suspends what they are doing or saying to hear what one person has to say. Do you know what I am talking about?

I've found that my dog Breck, regardless of what else is going on or who else is speaking, will listen to *my* voice. Through much training, Breck knows that when I speak, she'd better give me her eyes! I think about my dog when I read Psalm 123. The psalmist says, *"Unto thee lift I up mine eyes, O thou that dwellest in the* heavens. Behold, as the eyes of servants look unto the hand of their masters … so our eyes wait upon the LORD our God, until that he have mercy upon us." A servant looks to his master's hand for pretty much everything! Do you look to the Lord?

We should look to God in dependence on Him for our needs and for His direction in our lives. As Psalm 123 says, your eyes should *"wait upon the Lord."* My dog has eyes that wait on me. Ask any cowboy about the eyes of his cowpony, and he will tell you a useful horse has eyes that wait on the guy in the saddle.

Today, if life seems out of focus, it may be that you are looking to the wrong person. When you have a need, to whom do you look? Maybe you are looking too low! Look up to your Master. Strive to have "waiting eyes" that look to the Lord in every aspect and moment of your day.

◇◇◇

I'M SO BUSY!

DAY 139

Psalm 131:1 LORD, my heart is not haughty, nor mine eyes lofty: neither do I exercise myself in great matters, or in things too high for me.

In our modern society, we are so caught up in being busy that we often "bite off more than we can chew." Everyone is so busy these days! Being busy is almost a badge of honor, regardless of what you are accomplishing or what consequences may come. That is why I love Psalm 131. It gives perspective to people who do more than they can or should.

Have you ever stopped to think about all of the things that we can do more quickly than people could do a century ago? **Instead of having more time, we now can do more which leaves us *less* time!** When microwaves and modern conveniences came out, people believed that we could actually have more free time to spend with family. Instead of washing kids, clothes, and dishes, we can microwave a frozen dinner, eat it on the go, and stay out later. I'm not against modern conveniences, microwaves, or fast food, but these examples remind us of how busy we've become.

Haughtiness, or pride, is a terrible thing in the life of a believer. Pride is thinking more of yourself than you should. More specifically, **pride can be thinking more of your abilities than is actually true.** In other words, you think that you can accomplish more than you can. How much can you accomplish on your own? The Lord Jesus said, *"For without me ye can do nothing."* (John 15:5)

It is not virtuous to be lazy and never attempt anything great. It's not better to do less! The point is, biting off more than you can chew is a form of pride. **Humility is not thinking less of yourself; humility is not thinking of yourself at all.** Make the Lord your Source and follow Psalm 131:3, which says, *"Let Israel hope in the LORD from henceforth and for ever."*

What is it that you are trying today that is bigger than you? Don't live one second trying to accomplish things on your own! Instead, live in the calmness and quiet confidence (verse 2) that come with submission to God. Don't get so busy that you lose the proper perspective!

◇◇◇

DAY 140
GOD THINKS ON YOU; DO YOU THINK ON HIM?

Psalm 139:17 How precious also are thy thoughts unto me, O God! How great is the sum of them!

A guy calls you up and says, "Where are you?" You respond, "Who are you?" He says, "That doesn't matter. I know where you are." Three hours later he calls back and says, "What are you doing?" You reply, "Who is this?" He says, "No one of consequence; just tell me what you're doing. I *know* what you're doing." At the end of the day, before you drift off into sleep the same stranger calls back and says, "What are you thinking?" You reply, "I'm thinking that I'd be thrilled if you'd stop calling me." He says, "I know." That would be kind of creepy—*wouldn't it?*

Now, the feeling would totally change if the person calling were someone you loved. Suppose it was your grandma or your spouse or the person closest to you in all the world. They don't have to call you. You call them and say, "Guess where I am?" or "You'll never believe what I'm doing!" When you come home from work at the end of the day, your loved one is going to know what you're thinking whether he or she wants to or not. So the level of creepiness or comfort depends completely upon the person in question. If a stranger knows everything about you—what you think and who you are—that's ominous. If someone that loves you knows these things, it's a different matter altogether.

In Psalm 139:17, David reminds us that God thinks on us all the time, which is an amazing thing. He says, *"How precious also are thy thoughts unto me."* I think this implies both that God's thoughts should be precious to us and that His thoughts are often about us. The Psalm says, *"How great is the sum of them. If I should count them, they are more in number than the sand. When I awake I am still with thee."* God thinks about you, and He thinks about you all the time. The first few verses of Psalm 139 say that God searches, knows, and understands us.

So many times people don't know and understand us, but that is never the case with God. God takes a keen and active interest in us. Even when the people closest to us don't understand us, God does. It's an amazing thing that the God who has the universe to 'worry' about thinks about us constantly. The question is not if God thinks about you—He does. The question is how often you think about Him. Do you know where you are with Him? Do you know what He's doing in your life? Do you know what His thoughts are for your life? **God is a great God to think on me, and I would do well to think on Him.**

DAY 141

HOW MUCH CAN YOU ASK FOR?

Psalm 143:10 Teach me to do thy will; for thou art my God: thy spirit is good; lead me into the land of uprightness.

For how much can you ask God? For how much *do* you ask God? Psalm 143 teaches us that the more God owns, the more you can ask of Him. Now, God owns everything that we have or could ask for, but often we fail to see what God owns. **If you do not see God as owning everything that you need and are asking for, then you are not going to ask for what you should.** You can miss out on asking because you never see God as owning! The more God owns, the more you can ask of Him.

If you really needed something, would you rather ask someone with a lot of money and resources or some poor sap in the same condition as you are? Furthermore, would you rather ask a total stranger or a close friend? You are likely to get something much faster from your friend!

The point of Psalm 143 is not just getting "things"; the point is, **the more God owns *you*, the more you can ask!** The last five words of the psalm are *"for I am thy servant."* There is something wonderful about asking God for what you need because He can provide and He cares about you. Imagine if one of my boys asked a complete stranger for something that he needed. That is altogether different than his asking me!

Today, you can ask God for guidance in your life, but does He own your way? *"Cause me to know the way wherein I should walk; for I lift up my soul unto thee."* If your way doesn't belong to God, how can you expect God to show the way that you should go? If your finances don't belong to God, how can you expect God to provide your needs financially? If God doesn't own your family, how can you expect God to direct your children? The answer comes when you acknowledge that God owns *what* you need and that God owns *you*. **The beginning of God's answer is God's ownership.**

THE WAY TO KNOW YOURSELF IS TO KNOW GOD

DAY 142

Psalm 144:3 LORD, what is man, that thou takest knowledge of him! Or the son of man, that thou makest account of him!

There's nothing wrong with asking God to bless you or your life. But in Psalm 144, King David blesses or praises God. He praises God for who God made him to be. Psalm 144:1-2 says, *"Blessed be the Lord my strength, which teacheth my hands to war, and my fingers to fight; my goodness, and my fortress; my high tower, and my deliverer; my shield, and he in whom I trust; who subdueth my people under me."* This is a king who is blessing God for all that God is in his life and for all that he is because of God's working through his life. God was his strength and goodness. God was his high tower and shield. That's why David trusted in Him. God is a God who can be all these things to you today, as well.

But what is man to God? That's the same question David asks in the next verse. Verse 3 says, *"LORD, what is man, that thou takest knowledge of him! or the son of man that thou makest account of him!"* What is man? Verse 4 says that man is vanity, a breath, emptiness, a shadow that passes away. But verse 10 says that God is He that giveth. So when you have God you have all that you need; and until you come to a point where you realize your emptiness, you will not seek the God who fills needy people.

David ends the psalm by talking about the prosperity of the kingdom—healthy children, full crops, strong livestock—and wraps up by saying, *"Happy is that people, whose God is the LORD."* What made these people happy and blessed was not so much the things they had or attributes they shared, but it was the God they served who was the Provider of all these things. **You will never know what God made you to be until you know the God who made you.** So many times people are seeking to find themselves; and when they do, they find emptiness. What we need to do is to find God who is the Author of all that we need and all that we are not.

DAY 143

WHAT GOES AROUND ...

Proverbs 11:5-6 The righteousness of the perfect shall direct his way: but the wicked shall fall by his own wickedness. The righteousness of the upright shall deliver them: but transgressors shall be taken in their own naughtiness.

I visited a home several years ago in which there were boxes everywhere, stacked floor to ceiling. It was a sight to behold! At some point, the dear folks who lived there started collecting things which in time owned them.

Sometimes we do the same thing with trouble. We buy trouble that we could get for free. That is not smart! Proverbs 11 reminds us that **what you own today may own you tomorrow.** If you buy wickedness, you are going to own it; and at some point, it will own you. *"The wicked shall fall by his own wickedness ... transgressors shall be taken in their own naughtiness."*

In contrast, verse 17 says, *"The merciful man doeth good to his own soul...."* Do you want mercy in your life? If you give it, you will get it! That truth works in the positive and the negative. What you own today will own you tomorrow. Will it be wickedness, naughtiness, and cruelty, or will it be mercy?

If you look for trouble today, you will find it every time. Proverbs 11 is full of examples of the truth that what you own will own you. Which sounds better: being owned by evil and wickedness, or being owned by God's mercy and grace? As the saying goes, **what goes around comes around.** Will you spend your day buying trouble or buying goodness?

WHO ARE YOU WALKING WITH?

DAY 144

Proverbs 13:20 He that walketh with wise men shall be wise; but a companion of fools shall be destroyed.

In 1987, President Reagan was with Canada's Prime Minister Brian Mulroney at a hangar in Ottawa, Canada, awaiting the arrival of the motorcade carrying their wives. When the motorcade arrived and Mrs. Reagan and Mrs. Mulroney got out, President Reagan threw his arm around Prime Minister Mulroney, smiled broadly, and said, "You know, Brian, for two Irishmen, we sure married up."

That's a wonderful sentiment to have, and there's no doubting the fact that the people with whom we surround ourselves is vitally important. Proverbs 13:20 says, *"He that walketh with wise men shall be wise, but a companion of fools shall be destroyed."* There's a basic law of nature that you will be like the people with whom you surround yourself. You will go with and end up at the same destination as those who surround you.

Years ago, when a young person was in trouble at camp, my grandfather Bill Rice would often say, "You know, I can write that young man's story." What he meant was that he had seen this same story with different faces and names before. There was a certain amount of predictability about it because certain courses always end up at the same destination.

Now, I do need to be honest about the nature of my relationships with people. I should always be learning from some people, I should always be helping others, and I should always be aware which of the two I'm doing. People are important. They are the most important thing in the world because they are those for whom Jesus Christ died. And the kind of people we surround ourselves with is vitally important. **If you want to know what kind of path you're on, observe the people who are on it with you.**

DAY 145

WHERE ARE YOU HEADING?

Proverbs 14:12 There is a way which seemeth right unto a man, but the end thereof are the ways of death.

The Bible often speaks about direction and destination. Sometimes we can see one and forget the other, but direction and destination are tied together. One is right in front of you, while the other is off in the distance. However, if you are going to end up at the right spot, you have to go in the right direction! **The direction you are heading is important because it determines your destination.**

Any direction you may go will begin in your heart. Verse 14 says, *"A backslider in heart shall be filled with his own ways…."* The direction you are heading this morning is something that only you and God know. By the time someone else can see the direction you are heading, your heart is already at the destination. The journey to your destination **begins in your heart.**

Ask yourself this morning, "How am I trending?" Which way is your heart pointing? You will go the direction you are aimed. It is easy to have a slight bend toward living to the flesh or an attitude that is just slightly out of adjustment. You may think it is just a curiosity or interest, but the Bible goes to great lengths to warn us about taking heed to the direction we are heading.

Direction is important because it leads to a destination. You will end up in the wrong place when time and opportunity are applied to the wrong direction. That is why your attitude, the things you accept, the things you read, the things you watch, the things you listen to, and the things you think about are so vitally important. Let the Bible be your compass for your direction and destination. Where are you heading today?

THE VALUE OF YOUR WORDS

DAY 146

Ecclesiastes 5:1-2 Keep thy foot when thou goest to the house of God, and be more ready to hear than to give the sacrifice of fools: for they consider not that they do evil. Be not rash with they mouth, and let not thine heart be hasty to utter any thing before God: for God is in heaven, and thou upon earth: therefore let thy words be few.

Just yesterday I was reading Luciano Pavarotti's recollections of the 3 Tenors concert in Los Angeles, CA, in 1990. The stats are absolutely staggering. That concert had 56,000 people in attendance and was broadcasted to 107 different countries. It was a concert the magnitude of which is amazing. But **addressing God has more potential for power than addressing millions of people in word or song!**

Now, let me ask you a question. What are your words worth to the people who hear them? More importantly, how much are your words worth to God? Part of the answer to these questions is another question—How much is your word worth? Are your words true?

Scarcity and quality are two dynamics that add value to our words. Fewer words generally produce more value. **Our words are more valuable when we listen readily and speak deliberately.** This is true with our words before men and, more importantly, with our words before God. Here are a couple actions to keep in mind.

1. **Be ready to hear.** This verse is talking about our obedience. Don't just come to God with some empty ceremony or empty habit. Be ready to hear what God has said, and be ready to obey what you hear.

2. **Be slow to speak.** This is especially true when approaching God. We are not to be rash or rushed. Now he's not saying we should be reluctant to come to God in prayer. God is always

ready to hear. But our prayers should not be mindless or heartless, and our promises should not be empty.

In the end, the value of words is based on how true and beneficial they are. Anyone can have words that have value to God because they are true and they come on the heels of a listening ear and an obedient heart.

DAY 147 — WHY AND HOW

Isaiah 1:11 To what purpose is the multitude of your sacrifices unto me? saith the LORD: I am full of the burnt offerings of rams, and the fat of fed beasts; and I delight not in the blood of bullocks, or of lambs, or of he goats.

Have you ever forgotten something and run back into the room to grab it, only to completely forget what you were getting? You can make great strides but forget why! Israel was doing exactly that when Isaiah 1 begins. They were doing great things but had forgotten why.

God's people were doing several good things: sacrificing, coming into the courts of God, bringing offerings, and assembling together for religious reasons. God said, *"I am full of the burnt offerings … I delight not…."* He called their good works *vain, abomination, iniquity,* and *trouble.* He even said, *"When ye make many prayers, I will not hear…."* That is pretty strong! In short, the Israelites were doing all kinds of good things without remembering why they were to do them.

Knowing how to do something is important. A man who knows how will always have a place. But a man who knows *why* and *how* is much better! I have found that most of my sermon titles begin with *how* ("How to Pray," "How to Read Your Bible," "How to Be a Better Husband"). In fact, there is a whole host of books on "how to" do just about anything!

Ironically, as Christians, **we don't usually struggle with "how to"; we struggle with "why."** We know what to do and how to do it, but we forget why we are to do it in the first place! Be wise enough to remember why you are doing the good things you do. Nothing you do for God will please Him when you don't have a heart for God. Don't forget the why!

GOD'S SON IN GOD'S WORD

DAY 148

Isaiah 7:14 Therefore the Lord himself shall give you a sign; Behold, a virgin shall conceive, and bear a son, and shall call his name Immanuel.

To Whom does Isaiah 7:14 refer? Certainly, the ultimate fulfillment of this verse is found in a child, God the Son, who is literally "God with us." The Lord gave a sign to those in Isaiah's day, and Christ was the ultimate, physical fulfillment of God's written promise!

Like the people in Isaiah 7, can I encourage you to see God's Son in God's Word? The truth is, **if you do not see His Son in His Word, you are not likely to see Him in your life either.** How do you see God's Son? First, you see Him by observation. It is always good to think as you read the Bible. Have you ever found yourself thinking, after reading a passage, "Now what did that say?" My mind wanders easily! Pay attention to what the Bible says, and see Christ in the Bible by observation.

Second, you see God's Son in God's Word by acknowledging inspiration. That is, **God's Son and God's Word rise and fall together.** Matthew 1:21-23 quotes God's promise in Isaiah 7:14. Matthew wasn't giving his thoughts on what God might have meant; He gave exactly what God meant because God inspired both Isaiah and Matthew! You can't dismiss the Virgin Birth but embrace "God with us" at the same time.

Jesus Christ is God's Incarnate Word. The Bible is God's written Word. **God's Book is truly about God's Son.** If you will look for Christ in the Bible, you will find Christ in your life. The same Christ present at Creation and present one day at Armageddon is the same Christ mentioned in Isaiah 7:14, and you can take God at His word on that.

◇◇◇

DAY 149
ARE YOU LIKE THE DEVIL?

Isaiah 14:12-14 How art thou fallen from heaven, O Lucifer, son of the morning! How art thou cut down to the ground, which didst weaken the nations! For thou hast said in thine heart, I will ascend into heaven, I will exalt my throne above the stars of God: I will sit also upon the mount of the congregation, in the sides of the north: I will ascend above the heights of the clouds; I will be like the most High.

Isaiah 13:1 begins with "the burden of Babylon" in which God declares His judgment on the arrogance of that empire and its king. The Bible predicts that after this judgment people will look at the king of Babylon in awe that such a mighty one was now in such humility. Then right in the middle of this "burden," Isaiah 14:12-14 says, *"How art thou fallen from heaven, O Lucifer, son of the morning! How art thou cut down to the ground, which didst weaken the nations! For thou hast said in thine heart, I will ascend into heaven, I will exalt my throne above the stars of God: I will sit also upon the mount of the congregation, in the sides of the north: I will ascend above the heights of the clouds; I will be like the most High."*

Are these words speaking of the king of Babylon or are they speaking of the devil? Lucifer is a name that means "bright one," and yet the bright one is revealed to be a "five I-d monster": I will, I will, I, I, I. I'm never more like the devil than when I'm full of myself. The same is true of you as well. When your life is defined by worry, anger, or envy, it is because your life is defined, at that moment at least, by a focus on self. The more full of self we are the more like the devil we are because that was the quintessential, original sin—pride.

If you were to meet the devil on the street today, would you be able to identify him by his actions? What would those actions be? Would the action be something like selling drugs to a minor? Or could it just be the pride and arrogance that characterizes so many of even God's people?

Pride is an abomination to God, and pride is focusing on self when we should be focused on God. If I am focused on self, I can never see others clearly. When I'm focused on God, I can see both myself and those for whom Christ died just as clearly as is possible. In fact, I can see as clearly, in a sense, as God Himself because I'm seeing those people as God sees them. **I am never more like the devil than when I am full of myself.** And I am never more like the Lord Jesus than when I am full of the Spirit of God.

RECONCILED TO GOD, RECONCILED TO EACH OTHER

DAY 150

Isaiah 19:25 Whom the LORD of hosts shall bless, saying, Blessed be Egypt my people, and Assyria the work of my hands, and Israel mine inheritance.

Whether you watch the news or not, you are probably aware that war is rampant in our world. You're also aware that war is especially rampant around the nation of Israel. That has always been the case. Truthfully, there's war in our corner of the world too; and there is war in every human heart. That war is not primarily with other people, but with God Himself.

In Isaiah 19 the prophet looks ahead to a coming day when ancient enemies will be reconciled. They will be reconciled with each other because they will be reconciled with the God Who made them. The Bible says that there will be an altar to the Lord in the land of Egypt. The Egyptians will make sacrifices to God. There will also be roads between Egypt and Assyria because the Assyrians will be coming to Egypt to serve God with the

Egyptians. There will be unity among these nations who often have had one thing in common—their hatred of Israel.

However, the problem Egypt has now is the same problem Israel has: both have turned from God. Israel's problem is the same problem our country has today. We have turned from God. And the problem with these nations is the same problem with my own wicked heart and yours: we have turned from God. But, when those nations are reconciled to God, they will be reconciled to each other; and when we turn to God, we can be reconciled with others also.

Now let me ask you, with whom would you be reconciled today if you were reconciled to God? We war amongst ourselves because we war against God. You cannot honestly say, "Well, I'm right with God, but I hate somebody else." The Bible says in I John that if a man says he loves God but hates his brother, he's a liar. **When I give due diligence to be right with God, I cannot possibly live with malice towards anybody else.** When I make things right with God, I make things right with my world and the people in it.

DAY 151

PEACE THROUGH STRENGTH

Isaiah 27:5 Or let him take hold of my strength, that he may make peace with me; and he shall make peace with me.

God wants His people to be fruitful. That was true of Israel in Isaiah 27, and it is true for us today. God's character has not changed toward those He possesses. He provides in order to make you fruitful; His provision is not meant to end with you. Whatever God has given you, including your mind, talents, money, and time, He has given for a purpose.

Difficult times in life do not come because God hates you. In fact, difficult times come precisely because He loves you and is interested in what is best for you. Everything God allows in our lives is to make us fruitful.

When it comes to your life, **you will possess God's peace when you take hold of His strength.** What are you holding on to this morning? Often we look to what we can hold in our hands to give us strength. We might say, "If I just had more (blank), then I would…." Or perhaps we think, "If (blank) were different, then I would have peace." You can spend your entire day trying to build strength and security in your life, but you will have no peace. **Peace doesn't come through your strength; peace comes through God's strength.**

Christ is our peace (Ephesians 2:14); and when we take hold of Him, we can have His strength and peace. And this strength and peace can then be an encouragement to others. No matter what God is doing in your life, He desires that you be fruitful. Don't search for peace in all the wrong places; take hold of the One who is peace!

◇◇◇

ARE YOU LIBERAL?

DAY 152

Isaiah 32:8 But the liberal deviseth liberal things; and by liberal things shall he stand.

The first three words of this chapter are, *"Behold, a king…."* Isaiah 32 tells us about the perfect King, the perfect kingdom, and what life will be like in that kingdom. Verse 5 says, *"The vile person shall be no more called liberal, nor the churl said to be bountiful."* Now those are some fancy words! If you are like me, the word *liberal* does not strike a good chord; that is because today *liberal* is often used in a negative way to describe those who give away what does not belong to them. But the Bible word *liberal* means "generous." The idea comes from the word *voluntary*. In the sense intended here, **I am liberal when I volunteer what I have.**

We live in a day when evil is called "good," and good is called "evil." Terms mean something; and ironically, today a wicked man is sometimes called a liberal. A theological liberal is generous with the Word of God—but the Bible does not belong to him! A political liberal is generous with the United States Constitution (the founding fathers' words)—that is not theirs to give away either! The Bible says that there is coming a day when evil will be called what it is, and good will be called what it is.

But the Bible word liberal means…volunteer what I have. So what does this mean for us today? First, **we are known by our actions and our words.** That is sobering. Truly, we each have two names: the one our mother gave us and the one we are making for ourselves.

Secondly, **you will stand or fall by your actions.** Few things will come back to bless you more than your generosity—your volunteering—of self. Often we think of generosity as money (and it does include that), but being generous is so much more! What about your time, efforts, etc.? Are you known by your generosity?

What are you volunteering today? Are you volunteering your heart, efforts, help, and money? A person is truly generous when he is volunteering what he has. What will you volunteer today?

◇◇◇

DAY 153

DO YOU KNOW THE FUTURE?

Isaiah 33:5 The LORD is exalted; for he dwelleth on high: he hath filled Zion with judgment and righteousness.

Sometimes the future can be frightening because of what we do not know. What will today hold? What about this year? What about this lifetime? Well, those are tough questions to answer when you don't know the future! **Despite an unknown future, you can be confident because you can be confident about God.** I would much rather know God and not know the future that He has planned than to know the future in some measure but not have God.

Our society works hard at knowing the future. Amazingly, there is still a market for palm readers, horoscopes, and predictions! However, **what good is knowing the future if you do not know God?** Facing today without knowing the future is possible when you know the God of today and tomorrow.

The Bible says, *"The LORD is exalted; for he dwelleth on high...."* God is exalted and He dwells on high. In other words, **God has a "bird's-eye" view** of your life, your day, and your future. You don't know the future, but God does! Be reminded that God dwells in heaven, but He works on earth. Thank God that despite not knowing the future, we can know the One who does know the future and is exalted on high!

WHERE IS YOUR CONFIDENCE?

DAY 154

Isaiah 36:4 And Rabshakeh said unto them, Say ye now to Hezekiah, Thus saith the great king, the king of Assyria, What confidence is this wherein thou trustest?

Don't you just love sports interviews? Usually the conversations are not exactly "rocket science," but sometimes they are amusing to listen to because of all the "trash talking." The trash talking is not confined to sports stars; politicians who are campaigning can do a pretty good job too! In the Bible, we find an interesting passage about a wicked Assyrian who talked trash, but his story is worth our consideration when it comes to the subject of trusting God.

The king of Assyria sent Rabshakeh and a huge army to Hezekiah. His initial question is a thought-provoking one for us to consider: *"What confidence is this wherein thou trustest?"* To summarize his trash talk, he mocked Israel's trusting in Egypt

(v. 6), in Jehovah God (v. 7), and in Hezekiah (v. 14). By extension, he was trying to get the people of Israel to trust Assyria!

What is the object of your dependence? What is your *"confidence"*? Trusting in the world (Egypt) is pointless and empty. Trusting in yourself is not any better. Trusting in Jehovah God, although mocked by the world, is the only legitimate choice. How sad that a wicked Assyrian could not see the difference between trusting the Lord and trusting Egypt!

You can trust in yourself, you can trust in the world, you can trust in someone else, or you can trust in God. **Are you fully trusting God for everything you need today?** Would the world see you trusting the Lord, or would they see you seeking confidence from another source?

◇◇◇

DAY 155 — LET GOD SPEAK FOR HIMSELF

Isaiah 37:6 And Isaiah said unto them, Thus shall ye say unto your master, Thus saith the LORD, Be not afraid of the words that thou hast heard, wherewith the servants of the king of Assyria have blasphemed me.

Who has a greater capacity to cause you confusion or anxiety or trouble? Is it someone who obviously speaks for the devil or someone who claims to speak for God but, in actuality, speaks for the enemy? It would be fairly easy for many of us to dismiss a militant atheist or maybe even to have some level of pity on him. But what about a man who has been to seminary, a very educated person? What if he knows the Bible well, but he uses what he knows to make claims about God that stir confusion, anxiety, and trouble?

In Isaiah 36 Sennacharib, king of Assyria, laid siege to Jerusalem. His commander verbally attacked the Jewish people in two ways. First of all, he said, "Don't trust your God. No country's gods have been able to defend them to this point, and your God won't be the first." Now that was an obvious attack, but it was fairly easy for the people of Israel to rebuff. They easily saw his comments for what they were—an attack on their God.

The second kind of attack the commander gave was much craftier. In Isaiah 36:10 he basically says, "Look, I'm not here by myself. God sent me here. God is against you. And God is for me and has told me to come conquer you." He claimed to be speaking for God, and it was calculated to inflict fear and confusion in the hearts of Israel. What made this man dangerous was that he claimed to be speaking for God.

Here's the important truth: **Let God speak for Himself and beware of the words of a person who appeals to the authority of a God he does not know.** That means it's important for all of us to know what God has actually said. We hear many things every day. Some of them are discouraging and some are crafty, but all of them should be weighed in light of what God has actually said for Himself in His Word, the Bible. A people who know the Bible and ask God to help them understand it will not be easily misled.

◇◇◇

WHO ARE YOU SERVING?

DAY 156

Isaiah 44:18 They have not known nor understood: for he hath shut their eyes, that they cannot see; and their hearts, that they cannot understand.

I know when it is flu season because I get all sorts of information about flu shots. I recently received a card in the mail that could get me a discount on my flu shot. How miserable! I don't like to spend money on the flu! The flu is bad, and getting a shot is bad. If I had my way, I wouldn't touch either! No matter what you think about the flu shot, it is an inoculation against a coming menace. That is what Isaiah 44 is, an inoculation—a warning—from Almighty God against the wickedness of idolatry.

God ridicules the prospect of worshipping an idol. How ridiculous, He says, for a man to chop down a tree, use half of the wood to build a fire to cook his supper, and use the other half to build a god (verses 15-17). But the verse that caught my attention was verse 18: *"They have not known nor understood: for he hath shut their eyes, that they cannot see; and their hearts, that they cannot understand."* The person who depends on a manmade god that cannot see or hear is just as blind and without knowledge as his god!

This passage in Isaiah immediately reminded me of Psalm 115 and Psalm 135 in which idols had mouths, eyes, ears, noses, and feet but could not speak, see, hear, smell, or walk. Psalm 115:8 condemns the one who puts his dependence on them by saying, *"They that make them are like unto them; so is every one that trusteth in them."* A person who puts his dependence in such a god will go through life the same way—without perception, without knowledge, and without sight. In contrast, **God Almighty is everything you need for nourishment and knowledge.** Taking your hope from God and placing it in an idol is transparently shallow. What a shallow substitute for the real thing!

You are only as good as the God (or god) you serve. Is that good news or bad news? Well, that depends on the God (or god) you are serving! If my hope for the future is in some retirement fund or bank account (neither of which is wrong in itself), my life is only as good as the market. And how good is that right now? Living that way and trusting in what I know will only take me so far.

Maybe you are not trusting in some market account, but there is still a temptation to trust in some "god" that cannot provide. For instance, placing all my hope in my own savvy or my own abilities is a pretty shallow existence. I am only as good as the God (or god) I serve. If I serve a god that cannot hear, see, know, or perceive, then that is exactly the life I will have!

On the other hand, I honestly do want to know, see, and perceive what God wants me to know, see, and perceive. And the truth is, I can! I can if I am serving the right God. So whom are you serving this morning? To what (or whom) are you looking? Truly, you are only as able as the God (or god) you serve. Is that good news or bad news for you?

CLOD OR CLAY?

DAY 157

Isaiah 45:5 I am the LORD, and there is none else, there is no God beside me: I girded thee, though thou hast not known me:

Can you think of a clod—a very bad person—in your life right now? The term *clod* may be foreign to you; it is a lump of dirt or clay that is hardened and cannot be shaped. God often speaks of people in whom He is trying to work as "clay," so perhaps the term *clod* is appropriate!

In Isaiah 44:27-28, God speaks of a pagan king who was a clod. Cyrus was a pagan conqueror and a clod in every sense of the word. He was a terrible man. Yet everything Cyrus did for God's people he did as God's shepherd and for God's pleasure (verse 28). Amazingly, in Isaiah 45:1, Cyrus is called God's *"anointed"*! God says of Cyrus, *"Whose right hand I have holden...."*

What does all this mean for you and me this morning? The bottom line is that God is in control. He is the shaper of events, of times, of seasons, and of men. In light of that, **whether you choose to be clay (moldable) or a clod (dull and hardened), God <u>will</u> win.**

When President Reagan was asked some years ago what his strategy was for the Cold War, he replied, "We win. They lose." That is simple enough! By comparison, God's strategy is identical—He <u>will</u> win. He can win with you, or He can win despite you. It matters to me that He wins with me, not in spite of me. Doesn't it matter to you? I don't want to be a clod—I want to be clay!

Remember today that the clods are not in control. They may seem like they are in control sometimes, but remember that God is in control, not the clods. Isaiah 45:5 declares, *"I am the LORD, and there is none else, there is no God beside me: I girded thee, though thou hast not known me:"* God knew about

Cyrus long before Cyrus was even born! Remember, the clods are not in control.

Secondly, **remember that you don't have to be a clod!** God will work with you or against you, but God will win. As a child of God, why in the name of common sense would you want to be in opposition to Him? God will still have His way, even if I am not receptive and perceptive to what He is doing. You will be a part of what is happening one way or the other!

Thirdly, **remember to be clay.** Clay is pliable and moldable. Verse 9 says, *"Woe unto him that striveth with his Maker!.. Shall the clay say to him that fashioneth it, What makest thou?…"* Remember to be clay. I may not always know what God is making with my life, but God knows. There is a plan, a purpose, a design. God is looking for pliable people whom He can use and mold for His purposes. **The question then is simple. Are you a clod or clay?**

◇◇◇

DAY 158

HEAVY LIFTING

Isaiah 46:1 Bel boweth down, Nebo stoopeth, their idols were upon the beasts, and upon the cattle: your carriages were heavy loaden; they are a burden to the weary beast.

At the Grand Canyon, you will see pack mules carrying people and packages up and down the long, rough, rugged switchbacks. I think of those mules when I read Isaiah 46:1, which speaks of the Babylonians carrying out their heavy gods when they were invaded. I don't know how much an idol would have weighed, but I can imagine they were heavy if they were made of stone or wood. And these critters were the ones laboring to carry the gods out of a conquered city! Verse 2 calls these gods *"a burden."*

The idols had hands but could not touch; they had feet but could not walk. So instead of their gods carrying them, they were carrying their gods! **Instead of being a blessing, their gods were a burden.** How silly that these little gods could not deliver himself, let alone the people who worshipped him. A poor little donkey had to carry them out!

Contrast that with verse 3, which says, *"Hearken unto me, O house of Jacob, and all the remnant of the house of Israel, which are borne by me from the belly, which are carried from the womb."* God says, "I will carry you." God is up to the task, even when you are a little heavier and have a little more white hair than you did when you were born (verse 4).

Who is doing the heavy lifting in your life today? Has what you are trusting in become a burden? Living for God should not be a burden; if it is, take a look at who or what you are serving. God is the One who is able to carry us. Sometimes we can be serving, but our serving is self-serving. It is even possible to serve in a fine Christian ministry and not serve God. You are burdened with the weight of ministry because you are serving self. **Are you bearing the load instead of allowing God to bear your load?**

Our God will carry us, and serving Him is not a burden. I Peter 5:7 says, *"Casting all your care upon him...."* Why should we do this? *"For he careth for you."* God cares for you, and He carries you. In Isaiah 43:2 God says, *"When thou passest through the waters, I will be with thee; and through the rivers, they shall not overflow thee: when thou walkest through the fire, thou shalt not be burned; neither shall the flame kindle upon thee."* God is the One who will carry you; we need not rely on ourselves. How wonderful!

Our relationship with God does require maintenance, but it does not require carrying Him around. Fellowship with God requires time, thought, and maintenance, but <u>God</u> does not require maintenance! Are you carrying around a god today, or is God carrying you? Who is doing the heavy lifting in your life?

DAY 159

ALARMS

Isaiah 49:23b And thou shalt know that I am the LORD: for they shall not be ashamed that wait for me.

I don't know about you, but I live by alarms. If they don't go off, I am not where I am supposed to be! After I wake up to an alarm each morning, I check my email, the weather, and the things that will alarm me (quite literally) at some time throughout the day. Being late can be a very shameful thing!

In Isaiah 49, God is speaking to His people who will be taken captive and then later restored to the land. In verse 23, God speaks of a time when the tables will be turned—the conquerors will act as if they were conquered. But remember, this is written to people who would be in captivity their entire lives! God says, *"They shall not be ashamed that wait for me."* I love that! God wrote this promise to people who had not yet faced the captivity from which He was promising deliverance!

Truly, *"they shall not be ashamed that wait for [God]."* First, this means that **God is always on time.** He may not be on time according to my time, but His time is always the right time.

Secondly, this means that **God never forgets.** If I am late, usually this means I forgot or missed my alarm. God never needs an alarm—He never forgets! Israel thought that God had forgotten them (verse 14), but God compared His faithfulness to a mother and her newborn child. It would be more likely for a mother to forget her baby than for God to forget His children. A mother would never forget her baby, but God says, *"Yea, they may forget, yet will I not forget thee."* (verse 15) God will not forget!

Have you ever been waiting on something and wondered where God was? Maybe you wonder if God has forgotten or if perhaps He does not know your need. Well, be encouraged. You are looking at *your* watch—you are checking the wrong clock! God is always on time, and He does not forget. He reminded His people (and us today) in verse 16, *"Behold, I have graven thee upon the palms of my hands; thy walls are continually before me."* God knows and cares, and He is always on time.

A noble life is a life lived on God's watch and in God's timezone. There is no room for impatience, small visions, and small thinking in such a life. You don't have to wait on everybody, but you are wise to wait patiently on God.

"For they shall not be ashamed that wait for me."

◇◇◇

WHO ARE YOU? DAY 160

Isaiah 51:12 I, even I, am he that comforteth you: who art thou, that thou shouldest be afraid of a man that shall die, and of the son of man which shall be made as grass.

Any man with a sense of his own unworthiness may have a tendency to feel like he has no claim on God's peace or God's comfort. But God turns that notion on its head in verse 12: *"I, even I, am he that comforteth you: who art thou…."*

The Bible says that God's salvation and His righteousness are eternal—they go from generation to generation. In contrast, man is like the grass which withers in the hot sun or the garment that is eaten by moths (see verses 1-6). God says, "I am the One who comforts; who are you to fear?"

We say, "Oh, I have no right to peace or comfort or confidence," but God says, "Who are you?" The children of Israel had forgotten the LORD their maker (verse 13), and He reminded them that He is the One who made everything! God, the Creator of the universe, is the One who comforts—who in the world are you to live without His comfort?

If God made a nation from a barren couple (Abraham and Sarah—verse 2), who are you to fear? If God made the heavens and the earth and will destroy the same (verse 6), who are you to fear? If God made the law, who are you to fear?

Our tendency is to think that we are out of place to have peace and comfort. But if you belong to God, you are not out of place. In fact, **it is presumptuous to *not* receive what God is able to give.** In contrast to men who are like grass, God says, *"But I am the LORD thy God, that divided the sea, whose waves roared: The LORD of hosts is his name."* (verse 15) He is the LORD, and He is the "God of all comfort." (II Cor. 1:3)

God says, *"Who art thou?"* and the answer is simple: I am no one. **I have no right** to be fearful when God gives courage. **I have no right** to be doubtful when God gives confidence. **I have no right** to be discontent when God is the source of contentment. **I have no right** to live in lack when God is the One who provides.

"I, even I, am he that comforteth you: who art thou…."

◇◇◇

DAY 161
FOUR TRUTHS ABOUT YOU

Jeremiah 1:5 Before I formed thee in the belly I knew thee; and before thou camest forth out of the womb I sanctified thee, and I ordained thee a prophet unto the nations.

Living up to your potential means living up to God's purpose. Jeremiah was commissioned by God to preach a tough message to a tough group of people. Because of his own abilities and the circumstances surrounding him, Jeremiah was self-defeated. He said, *"Ah, Lord GOD! behold, I cannot speak: for I am a child."*

Haven't we all felt like Jeremiah at some point? "Lord, I cannot." **The task needs not match our strength; our strength needs to match the task.** The Lord encouraged Jeremiah with four statements which should also encourage us to live up to God's purpose.

First, God said, *"I formed thee."* Whoever you are, God made you. Your fingerprints, your chemistry, and your genetics are unique to you. You were formed and designed by God Almighty!

Second, God said, *"I knew thee."* God knew you *before* He formed you. Even though sometimes people do not know you, God knows you. You were in God's heart before you were on this earth.

Third, God said, *"I sanctified thee."* You are different by design and on purpose. God has set you apart. All of us can be lured in by the mainstream; but the truth is, you *are* different. The question is not, "Should I be different?" You *are* different! The question should be, "Am I showing that I am different?"

Fourth, God said, *"I ordained thee."* God not only set you apart, He also set you on a course. You may not know everything about your course, but God does. Five steps down the line may not be clear to you; but if you will take the first step in front of you, you will end up where you ought to be five steps down the line.

Living up to your potential this day and this year simply means living up to God's purpose. **The only way to live up to God's purpose is to surrender to God's ownership.** He made you; He knows you; He set you apart; and He has a direction for your life. Live today in light of these four truths, and you will live up to God's purpose for you.

NO SENSE IN SENSES

DAY 162

Jeremiah 5:21 Hear now this, O foolish people, and without understanding; which have eyes, and see not; which have ears, and hear not:

Have you ever known someone who had no sense? No, not that kind of sense! Have you ever known someone who had lost their sense of hearing, smell, taste, touch, or sight? There are some things in life for which you need your five senses; but **if you live life depending upon your senses, you will not have any sense!**

Some issues and facts of life cannot be discerned by your senses. For instance, you cannot discern right and wrong by your senses. A classic response of someone living by their senses is, "I don't see what is wrong with that!" or, "I don't feel that it is wrong." You cannot base right and wrong on how you feel, what you see, or how something sounds! Your five senses are not sufficient to let you know what is morally right and wrong.

My chocolate Labrador, Breck, can see, taste, hear, touch, and smell; but I would not trust her to discern what is best or right for me. Some Christians are no better off than I would be if I trusted my dog!

Jeremiah 5 gives us a better way to live. The right way to live is with a "sixth sense" from God and His Word. A Spirit-filled heart that is guided by God's Word is a greater "sense" than your sight, smell, taste, or touch. If your heart is not inclined to hear, you will not hear. If your heart is not inclined to see, you will not see. **The only way to clearly see and hear is to have a right heart in tune with God!**

Your greatest need this morning is to know God's mind on any given matter you face. Sometimes that may mean not trusting your other senses. Like a pilot that flies by the plane's instruments, not his own sight, may God help each of us to have the sense to live by His Spirit and His Word.

FORWARD OR BACKWARD?

DAY 163

Jeremiah 7:24 But they hearkened not, nor inclined their ear, but walked in the counsels and in the imagination of their evil heart, and went backward, and not forward.

Proximity to God is based on what you *are*, not what you *do*. It is easy to think that if you do a certain list of things, you will be fine. Serving the Lord and being part of a good church are good things to do, but doing good things can never replace a right relationship with God.

It is much easier to judge yourself by a visible, fixed standard of rules and regulations to see if you are square with God. For example, think of any standard we often make for a "good" Christian: reading the Bible, going to church, or going soul-winning. All of those things are good, and you should do all of them, but they are not the standard! You can read the Bible, go to church, and go on your church's visitation program, yet still have a wicked heart. The standard for being right with God is not hitting a set of marks; **the standard is the direction I am heading.** You can do right things while pointed in the wrong direction!

Based upon your relationship to God, how are you doing today compared to last year at this time? **Are you trending closer to God, or are you going backward?** The question is not whether you are following a set standard of orthodoxy or doing good deeds. In relation to God, are you going forward or backward? Take heed to the example of the children of Israel. They were in God's house, doing good things, yet they *"went backward, and not forward."*

DAY 164

PROPER PERSPECTIVE

Jeremiah 9:23-24 Thus saith the LORD, Let not the wise man glory in his wisdom, neither let the mighty man glory in his might, let not the rich man glory in his riches: But let him that glorieth glory in this, that he understandeth and knoweth me, that I am the LORD....

A clear view of God will give you a good perspective of yourself. While you can compare yourself with others and exalt yourself above those around you, you cannot continue in pride when you accurately see God in His place!

I remember as a youngster getting a "behind the scenes" tour of the White House from a friend who protected the Vice President. We went through a back gate, in a back door, and through some of the notable rooms in the White House. I recall seeing the Presidential barber shop, cabinet meeting rooms, and the press room; but the room that stands out in my mind is the Oval Office. We were not able to go into the Oval Office, but I distinctly remember the western artwork and the Bronco Buster statue in President Reagan's office. I thought it was cool that the President was a cowboy!

The White House is nice and all, but did you know that I have an office too? In fact, my office is bigger than anyone else's that I know. I even have western artwork and a cheap imitation statue of Remington's Bronco Buster, but does my office impress people like the Oval Office does? Only in my dreams!

Comparing the Oval Office to my office at the Ranch seems silly because of our perspective. I might be able to compare my office to yours; but when you see the Oval Office, there is no comparison. **In the same way, knowing God is the key to an accurate estimation of self.** It is not what you know; it is not what you do or how you do it; it is how well you know God.

God is more concerned with Christlike character than He is with man-produced talent of some sort. Whatever ability you possess was given to you by God, and the highest goal of everything you do should be to know God better. With a clear view of God, you will have a proper perspective of self.

NO UPGRADE REQUIRED

DAY 165

Jeremiah 10:10 But the LORD is the true God, he is the living God, and an everlasting king: at his wrath the earth shall tremble, and the nations shall not be able to abide his indignation.

Do you have a computer more than five years old? Do you drive a vehicle that is more than ten years old? Do you have clothes that are more than three years old? If so, what are you probably interested in doing? Upgrading! I must admit, I recently realized that I have a suit that I bought nine years ago. It was pretty stylish back then, but I'm not sure it is in style today!

The truth is, anything man-made needs to be upgraded and improved at some point, whether that involves computers, cars, clothes, or a god of someone's making. Jeremiah 10 reminds us that a man-made god always needs an upgrade. A god that you can make is no more able to care for you than a computer, car, or piece of clothing. **You can only improve on a god of your own making.**

In contrast, there is none like God. He is not man-made; He made man. **He is exclusively and utterly unique, and He never needs improving!** He is *"the true God, he is the living God, and an everlasting king."* He is *"the former of all things."* You cannot and will not improve on God.

We can get so lost in this world's goods that we forget that anything man-made will always need upgrading and improving. We think, "If I just had that, then I would be happy." God's advice in Jeremiah is, *"Learn not the way of the heathen, and be not dismayed at the signs of heaven; for the heathen are dismayed at them."*

God is not fashionable, trendy, or popular; God is *timeless*. **You cannot improve upon or do better than the God who made you.** You can only improve a god of your own making, and any god you can make is not worth worshipping. Follow the true God, the living God, the everlasting King.

◇◇◇

DAY 166 — THE FOLLY OF REJECTING GOD'S WORD

Jeremiah 36:23 And it came to pass, that when Jehudi had read three or four leaves, he cut it with the penknife, and cast it into the fire that was on the hearth, until all the roll was consumed in the fire that was on the hearth.

Imagine if the prophet Jeremiah were in your city or town this morning. What if he actually read aloud the very words that he received directly from God? The very words of God came from the mouth of His servant! You might be tempted to think, "I would get more from God's Word that way. I mean, straight from the source!"

However, Jeremiah is not going to appear in front of you today. By neglecting the Holy Spirit's guidance with God's Word today, you can end up just like King Jehoiakim. He sliced and diced the Word of God, then threw it in the fire. "Oh, Brother Wil," you say, "I would never do that!" **To disregard God's Word and His Spirit is little better than to willfully destroy God's Word!**

The truth is, you can get as much from God's Word as you would get if the penman were standing before you. On the other hand, you are rejecting God's Word just as much as Jehoiakim did when you do not listen to what the Bible says. You may never rip up your Bible and burn it; but when you hear preaching, do you neglect what God is saying to you through His Word? Do you reject Bible truth that doesn't appeal to you?

You cannot destroy the Word of God. When you try to destroy His Word, you end up destroying yourself. Whether you are cutting out parts of it, burning it, or simply disregarding it, you will not destroy God's Word. God had His way with Jehoiakim and commissioned Jeremiah to write the Word of God again. God's Word is eternally settled in heaven, and nothing you can do on this earth will change the fact that it's settled!

We stand in jeopardy if, for any reason, we are missing what God is saying. Reading or hearing the Bible than ignoring it or rejecting it altogether is a dangerous thing!

◇◇◇

WE UNDERSTAND WHAT WE WANT TO

DAY 167

Jeremiah 38:15 Then Jeremiah said unto Zedekiah, If I declare it unto thee, wilt thou not surely put me to death? And if I give thee counsel, wilt thou not hearken unto me?

Jeremiah was a man who feared God. Jeremiah feared God more than he feared people. He obeyed God instead of cowering to the king. Because of this obedience, Jeremiah understood much more than just about anyone else did. It wasn't that he was smarter, but he was more willing to act on what he knew. Sometimes we don't know something because we don't want to know it. Sometimes we don't understand something because admitting that we do understand would cost us too much.

King Zedekiah, on the other hand, never seemed to get it. His lack of understanding was not because he hadn't heard but because it cost him too much to admit that he understood.

For instance, in Jeremiah 38, a group of princes came to him and said, "Look, this guy Jeremiah is not patriotic. He's telling us that judgment is coming from the Babylonians. He ought to be killed." To which the king replied, "Hey, I'm just the king, I can't tell you no. If you want to kill Jeremiah, then have at it."

He was afraid of what people thought. So, these wicked princes threw Jeremiah into prison.

Just after this, an Ethiopian eunuch came to him and said, "Hey, this is evil what these men have done. This isn't right." Zedekiah replied, "Boy, you're right. Let's let Jeremiah out of prison." So he did wrong when it was easy, and he did right only when there was pressure on him to do right.

Finally, he told Jeremiah, "Jeremiah, I've got to hear from God. Now, don't tell anyone what you're doing. I'll meet you at the third door of God's house." The third door was a kind of back entrance. You get the sense that this was a secret meeting. Then the king said to Jeremiah, "I will ask thee a thing; hide nothing from me."(Those were ironic words coming from a king who was trying to hide everything!) The meeting itself was a secret because this king was more concerned about what people thought than what God knew. As a consequence, he never understood what God would have him to understand.

Zedekiah had a hard life because he was more fearful of what people thought than what God knew. The lesson to be learned today is this: **It is impossible for a person worried over the favor of men to understand the words of God.** When you want to understand, and you determine to act upon what you know: you will have god's grace in your life.

◇◇◇

DAY 168
WILLING TO HEAR "NO" AND SAY "YES"

Jeremiah 42:2b-3 And pray for us unto the LORD thy God, even for all this remnant; (for we are left but a few of many, as thine eyes do behold us:) That the LORD thy God may shew us the way wherein we may walk, and the thing that we may do.

Have you ever had someone ask your advice when they didn't really want it? Few things are more insulting than to have people ask your advice when they don't plan on following it. They don't really care what you have to say. The only thing worse is your child asking for permission when really all he wants is a rubber stamp on what he already intends to do.

That was the attitude of the remnant that came to Jeremiah and said, "Ask God what He wants us to do and where He wants us to go." Now, those are worthy requests. If they're an honest reflection of a sincere heart, they are requests that God will always answer. However, when my mind is set on doing one thing but I'm asking just to humor God, the question becomes a sham. That is worse than not asking at all. God is moved, not by my asking but by my willingness to act on what He says.

Jeremiah replies to the remnant in verse 20, *"For ye dissembled in your hearts, when ye sent me unto the LORD your God, saying, Pray for us unto the LORD our God; and according unto all that the LORD our God shall say, so declare unto us, and we will do it."* He goes on to essentially say, "Look, you've already decided in your hearts what it is you want to do."

Israel had a go-to source in times of famine, war, and distress—Egypt. They automatically (and ironically) went back to their former slave masters to the exclusion of the God who loved them and made them. So basically Jeremiah said, "You're asking a question for which there is no answer because no matter what God says, you're going to do your own thing."

Here's the point: God is good. God knows you. God made you. **God will always give you as much direction as you need when you are willing to act upon it.** To know the way to walk and the thing to do, you need to be willing to hear 'No' and be willing to say 'Yes' to God. Don't ask God if you're unwilling to obey God. God will never withhold from you anything you need to know in the time that you need to know it when you're willing to act upon it.

DAY 169

SELF-SERVING

Jeremiah 45:5 And seekest thou great things for thyself? seek them not: for, behold, I will bring evil upon all flesh, saith the LORD: but thy life will I give unto thee for a prey in all places whither thou goest.

I love technology. With just a few clicks on the mouse and a tap of some keys, your work is *d-o-n-e*. Did I mention that I hate technology? With just a few clicks on the mouse and an accidental tap of some keys, all of your hard work can be *g-o-n-e*! Do you know the feeling? Now imagine if you had handwritten every word, only to have the king rip up your scroll and throw it in a fire! Baruch, Jeremiah's scribe, was serving the Lord and doing right, yet everything he had done was destroyed and unappreciated.

Baruch was discouraged about the work he had done for the Lord. His God-given service to write the words of God had become *his* work instead of the *Lord's* work. He felt sorry for himself and said, *"Woe is me now! for the LORD hath added grief to my sorrow...."* He was one discouraged servant of God!

The truth we can learn is, **if you work for yourself, you are more likely to feel sorry for yourself.** It is possible to serve yourself; to serve God; or to serve God for yourself. The danger for you this morning may not be to selfishly serve yourself. The greater danger may be to serve God for yourself.

The way we respond to adversity and down times can reveal our faith, or lack thereof, in God's ability to perform what He has said. It is *God's* work, after all! Serving the Lord requires stewardship, but even the best steward can fall into the trap of feeling sorry for himself. Be a good steward with what God has entrusted, but don't allow adversity to lead you to feel sorry for yourself.

ONE BOMB TARGET

DAY 170

Jeremiah 48:29 We have heard the pride of Moab, (he is exceeding proud) his loftiness, and his arrogancy, and his pride, and the haughtiness of his heart.

I have recently heard the term "one bomb nation" in reference to nuclear weapons. A "one bomb nation" is a country whose size makes being totally wiped out by one nuclear weapon a possibility. While current issues make this a sobering thought for today's world, a nation in the Bible was actually destroyed by one single weapon centuries ago. On top of that, this destruction was an inside job! They employed this weapon against themselves. No one did this to them; it was destruction of their own doing!

The nation of Moab destroyed themselves with the weapon of pride. How astonishing that one nation was conquered and later destroyed over the sin of pride! They had a plan against every type of weapon and enemy of their day, but their destruction was an inside job. A mighty nation was brought to nothing because of pride, loftiness, arrogance, and haughtiness!

The same weapon that destroyed Moab will be deployed by Satan upon a nation, a church, a family, or a person. We are a "one bomb target." Most of the sin and trouble you hear about can be traced back to pride. Pride has brought down nations, and it can bring down a Christian like you! **Pride gives life to sin but death to the sinner.**

Can you imagine how different your life would be if you did not have to always be first? Most of the trouble I get myself into comes when I insist on always being first, always being right, always being pleased, and always being recognized. Anytime I magnify myself and the things I have done, I am ultimately magnifying myself against the Lord (verse 26).

You are a target for the pride bomb. It will destroy your life and set you at odds with God (James 4:6). May God help us to have nothing to do with the sin of pride today!

"Pride goeth before destruction, and an haughty spirit before a fall."

DAY 171

GOD'S PERSPECTIVE

Jeremiah 51:17 Every man is brutish by his knowledge; every founder is confounded by the graven image: for his molten image is falsehood, and there is no breath in them.

When I read Jeremiah 51, I am struck by the fact that God is speaking through one man, Jeremiah, and addresses kings, kingdoms, and empires. He talks of those of antiquity, those of the present, and those yet unborn. He speaks of the past and of prophecies as if they occurred on the same day. Realizing that God has a perspective that exceeds my own lends a little serenity to my day.

God's perspective exceeds my own in the matter of time. Jeremiah 51:17 says, *"Every man is brutish by his knowledge; every founder is confounded by the graven image: for his molten image is falsehood and there is no breath in them."* Compared to God and His knowledge and wisdom, man is like a beast. There's so much man does not know!

Jeremiah 51:19 speaks of God being before all things. He is timeless. God spoke the world into existence at the beginning, and He also waits in the future. While we fuss and worry about timing, God is a master of timing. Sometimes we think God is taking too long. Sometimes we think God acts too quickly. We think these things because we don't have God's perspective. Trust God's perspective regarding time.

God also has a superior perspective when it comes to the purpose of life. In verse 20 He speaks to the conqueror and says, *"Thou art my battle ax and weapons of war."* Though our purpose is often short-sighted, God has a grand perspective. God has given us free will; but at the end of the day, God reigns. Conquerors like Nebuchadnezzar and Cyrus may have thought that their purpose was to rule the world, but God ruled them. They were nothing but tools in His hands. That's important to remember about ourselves and those around us. God is in control. He has a different perspective regarding time and purpose.

Lastly, **God has a different perspective on reward.** We can easily think that the evil prosper and the righteous suffer; but the truth is, God is a God of recompense. In the last part of verse 33 God says of Babylon, *"Yet a little while, and the time of her harvest shall come."* And again of Israel in verse 5, *"For Israel hath not been forsaken, nor Judah of his God."* So whether it be for good or for evil, God is the righteous Judge of all creation. There is much you can't see today, and that may worry and hurry you. But remember that God is neither worried nor hurried, and you can trust Him for your day, your life, and your eternity.

◇◇◇

MAKE A LIFE

DAY 172

Jeremiah 52:34 And for his diet, there was a continual diet given him of the king of Babylon, every day a portion until the day of his death, all the days of his life.

David Davies was a life-long inmate in the British prison system. He was, in fact, a rather gentle man and ideal prisoner, coming to be known as the "Dartmoor Shepherd." While serving a sentence for petty theft of a church alms box, he met the Home Secretary of Britain, Winston Churchill. Mr. Churchill worked to have Davies released since his offenses were minor and his behavior was a model for others. But alas, Mr. Davies' freedom lasted only one night! Once a free man, he resorted to petty crimes and was arrested again. David Davies lived out his remaining days in a prison cell in Britain.

That story may sound strange to you because you value freedom. However, Mr. Davies valued the stability of prison! This same story is repeated often in our society today. As hard as we try, **we Christians tend to value stability and slavery over freedom and faith as well.** The book of Jeremiah ends with the sad account of King Jehoiachin living out his days in slavery to the king of Babylon. Jehoiachin received a high-ranking job, a clothing allowance, and warm meals every day! Sounds like a sweet deal, right? The only catch was, *he was still a slave in the prison of the king of Babylon.*

Jehoiachin was making a living, but he was not making a life. There is a huge difference! Making a life requires living by faith and trusting God to provide your needs, even when you cannot see the way. John Rice used to say that he lived a "hand to mouth" existence: God's hand to his mouth. The alternative to freedom and faith is certainty that is guaranteed by slavery. Think about the times that the children of Israel wanted to go back to slavery in Egypt instead of living in God's special daily provision!

Making a living and making a life are two different ideas. Living in freedom requires living by faith. Although you might sometimes face the unknown, uncertainty with freedom trumps certainty with slavery any day. Don't spend your day making a living; make a life by faith in the God who can provide everything you need.

◇◇◇

DAY 173
DIFFICULTY BRINGS CLARITY

Lamentations 1:7 Jerusalem remembered in the days of her affliction and of her miseries all her pleasant things that she had in the days of old, when her people fell into the hand of the enemy, and none did help her: the adversaries saw her, and did mock at her sabbaths.

I've read stories of people who have fallen off the subway platform in Manhattan onto the tracks several feet below. In a couple of cases they fell just as a train was quickly approaching the station platform. In each story a stranger, at great risk to himself, jumped off the platform to rescue the person down on the tracks.

A person finding himself in this terrible situation would not care that he is directly under Manhattan, the heart of the financial capital of the world. This doesn't matter when his life is in jeopardy; he just wants to live. He's just grateful for a stranger who is a true friend by helping in his moment of need.

Hard times have a way of clarifying what is really important in life and who is really a friend. The people of God in Jerusalem found this out the hard way. The book of Lamentations describes their suffering after the fall to Babylon. The first couple of verses show how their misery was compounded by the memory of what they once had.

Jerusalem discovered what was important to her. Verse 11 describes how the people gave their riches for food to eat. It's amazing how our perspective changes when we stop to consider what is truly important! If some catastrophe occurred tomorrow, I think we'd be more grateful for the things we take for granted today: food, shelter, warmth. These things are all blessings, but we rarely think about them unless times are difficult.

Jerusalem also clearly saw who her friends really were. Verse 8 says that all who once honored her now despised her. Verse 29 says that Jerusalem called for her lovers, but they deceived her. Verse 21 says, *"There is none to comfort me."* A friend is not merely someone who celebrates with us in great and happy times, but someone who helps us when times aren't so great.

Each of us are keenly aware of what is important and who our friends are in a time of calamity. Why not consider and thank God for those things and those people today? Why not ask God to help you find a person in a dark spot and be that friend to them today?

DAY 174

SUNNY OR MOSTLY CLOUDY?

Lamentations 2:1 How hath the Lord covered the daughter of Zion with a cloud in his anger, and cast down from heaven unto the earth the beauty of Israel, and remembered not his footstool in the day of his anger!

Isn't it amazing how much better you can feel when the sun is shining? The world seems like a different place when there is abundance of sunshine, but what exactly does the sun provide? Well, it provides warmth, light, and energy. The truth is, the sun's shining helps me to shine! Yes, the sun actually helps or hurts my mood.

All of those things—warmth, light, energy, and a cheery mood—can be gone in an instant when something comes between you and the sun. The sun doesn't stop providing warmth, light, energy, and joy, but you may not see or feel its effects when it is cloudy. Have you ever felt like there was a cloud between you and the warmth and joy of God? That is exactly where Israel found herself in Lamentations 2!

You can put up with any cloud as long as the cloud is not between you and God. You cannot afford to ever live that way! God's presence provides light and fellowship. I John 1:5 says, *"…God is light, and in him is no darkness at all."* If you are living out of fellowship with God, you are living under a cloud. There is no warmth, no light, and no joy under that cloud!

On the other hand, Psalm 80:3 reminds us that **you do not need happy circumstances if you have the smile of God.** *"Turn us again, O God of hosts, and cause thy face to shine; and we shall be saved."* The best of circumstances mean absolutely nothing without the face of God shining on you. The warmth, power, and joy of God are just a choice away. Living in God's light is a choice you can make, no matter what clouds may blow your way. Is today's forecast for your fellowship with God sunny or mostly cloudy?

OWNERSHIP EQUALS OBLIGATION

DAY 175

Lamentations 2:20a Behold, O LORD; and consider to whom thou hast done this.

Jeremiah was a sensitive man. He looked around him and was dismayed by the sin of Judah. They had rejected God's message, and God's judgment upon them was severe. You can hear the sorrow in his voice when in Lamentations 2:20 he says, "Behold, O LORD, and consider to whom thou hast done this." God had meted out destruction on His own people.

Why would God deal so severely with His own people? God was keeping His own promise that was based on His own character. Verse 17 says, *"The LORD hath done that which he had devised."* Verse 8 says, *"The LORD hath purposed to destroy the wall of the daughter of Zion."* He did what He said He would do. God's people had sinned, so He had to deal severely with them.

In verses 6 and 7 note how many times the possessive pronoun is used in reference to God. Verse 6 says, "*He* hath violently taken away *his* tabernacle…*he* hath destroyed *his* places of the assembly." Verse 7 says, "*He* hath abhorred *his* sanctuary." In other words, one of the reasons God dealt so severely with these people was precisely because they were *His* people. He hated their sin; He didn't hate them. Even though He hated their sin, He loved them; and because they were His, He had to deal with them.

The overarching truth is that **ownership equals obligation**. God's ownership of me obliges Him to deal with me in light of His character and my need. The Bible tells us that God chastens us as dear children. My ownership of what God has put in my hand equals obligation as well. God had given these precious people a king, princes, a law, prophets, and a temple. Because they had so misused what God had given them, God took the gifts away and gave judgment instead.

How we use what God gives us determines what God gives us. I don't know what you own or are entrusted with today, but remember that God owns you. My ownership equals obligation, and God's ownership equals obligation. I need to remember that how I treat what God has entrusted to me does, in measure, determine how God will deal with me. I am His, and He is not indifferent towards me.

DAY 176
GOD'S MERCY IS AS FRESH AS OUR NEED

Lamentations 3:22-23 It is of the LORD's mercies that we are not consumed, because his compassions fail not. They are new every morning: great is thy faithfulness.

Some things are only good fresh. You're almost never going to see the word *fish* on a chalkboard outside a restaurant without the word *fresh* preceding it. We think about fresh flowers. Flowers are beautiful, but they certainly don't last forever. A wilted flower is worse than no flower at all. Or maybe the only thing your house needs is a coat of fresh paint. You know, paint is great, but paint when it's faded and cracked is almost more an insult than a help. How good then to know that the mercy that God offers to you today is of the fresh variety.

Lamentations 3:22-23 says, *"It is of the LORD's mercies that we are not consumed, because his compassions fail not. They are new every morning: great is thy faithfulness."* You're not faithful, but God is. You're certainly not perfect, but God is. The kicker is that God is both perfect and merciful. The lovingkindness that God extended to His wayward people so long ago is the character of God that has not changed even today. The same God that gave mercy and help to His people thousands of years ago is just as new, available, and fresh to a modern person in a modern world this very moment.

The morning never dawns on God because God never sleeps. He sees the night as if it were day. There's not a morning that dawns that can surprise or dismay God. Now it may be that today the sun is behind the clouds, and it may be that your life seems to be under a cloud. That may be true, but it's not unusual.

Most of Lamentations 3 paints a dark, dark picture. In fact, Jeremiah said in verse 2, *"He hath led me, and brought me into darkness, but not into light."* He goes on to describe the judgment of a righteous God on an unfaithful, wicked people. He was even to the point that his hope had perished. But hope sprang to life when he recalled in his mind God's mercy! Wherever you've been, wherever you are, and whatever the future brings, please know that **God's mercy is as fresh as your need.**

WHAT IS SUCCESS?

DAY 177

Ezekiel 2:7-8 And thou shalt speak my words unto them, whether they will hear, or whether they will forbear: for they are most rebellious. But thou, son of man, hear what I say unto thee; Be not thou rebellious like that rebellious house: open thy mouth, and eat that I give thee.

All of us want to succeed. But what is success? If you haven't defined success, then how are you ever going to reach it? Instead, you will meander through life until the end when you'll look back and realize that nothing happened of eternal value.

Perhaps the only thing worse than not correctly defining success is wrongly defining it. Some very ambitious people are going headlong toward the wrong goal. If the devil can deceive you about success, he has won before he has even begun.

So what is success? That would be a good question to ask the prophet Ezekiel. Ezekiel had been surrounded by loss. When his nation went up against the nation of Babylon, his nation lost. He lost his home and possessions, and was in captivity in a land far away. His entire life at this point had been defined by loss. He must have been hungry for success.

The message God gave Ezekiel for rebellious Israel was not a pleasant one. Imagine how Ezekiel must have felt when God came to him and said, "Look, the message I have for you to deliver to Israel is not a happy message, and the people who will hear it from you are not going to be very happy. In fact, I won't promise you that they will even accept your message. But don't be afraid of them." Would Ezekiel have felt successful?

If you look for a definition of success in Ezekiel 2, you come away with a very clear understanding that in God's eyes **success is obeying God and leaving the results to Him**. I'm not saying the results don't matter. What I'm saying is that, ultimately, we can accomplish many things but miss success. Whatever you do today, remember that success is obeying God and leaving the results to Him.

◇◇◇

DAY 178
ARE YOU HARD-HEADED?

Ezekiel 3:9 As an adamant harder than flint have I made thy forehead: fear them not, neither be dismayed at their looks, though they be a rebellious house.

We are living in a day where broad-mindedness is considered to be a great virtue. Ironically, being broad-minded today means being closed-minded to the truth. Often "broad-minded" is defined by those who are very closed-minded to God and the truth! Is it possible to be open-minded and hard-headed at the same time? Yes! Actually, it is imperative that you are!

The only way to be open-minded to the truth is to be hard-headed against anything that strays from the truth as defined by God. Ezekiel had to be hard-headed with God's difficult message to a bunch of hard-hearted people. He was to be hard-headed in his pursuit of the truth and open-hearted to what God had to say. One side demands the other. You can only be open-minded to God's truth when you are hard-headed *about* God's truth.

When you have an open mind toward God, and a heart for the people He loves, it is reasonable to be a little hard-headed for God too. Don't let having a broad mind equal having an empty head and heart. To lock God's truth in, all falsehood must be locked out. "If you stand for nothing, you will fall for anything."

The only truly open mind is a mind open to the truth. The only way to be open-minded to the truth is to define it by absolute authority and be hard-headed against anything that strays from God's standard.

◇◇◇

DISTRACTED DRIVING
DAY 179

Ezekiel 8:16 And he brought me into the inner court of the LORD'S house, and, behold, at the door of the temple … were about five and twenty men, with their backs toward the temple of the LORD, and their faces toward the east; and they worshipped the sun toward the east.

Have you ever found yourself driving someone else's car? I don't mean you were actually behind the wheel in the driver's seat; I mean the other driver was making you nervous or impatient by the way he was driving. The other night on the way to church, I noticed a fellow driver who made me nervous and frustrated at the same time! He was completely distracted, reaching for things on the floorboard and overhead. He was looking everywhere and at everything except the road! Consequently, his car was weaving in and out of both lanes while speeding up and slowing down for no logical reason. The worst part was, he was doing all of this while camped out in the left lane of the interstate that is supposed to be for passing!

I must confess that I am very impatient with other drivers like the one I've just described. However, sometimes I am that driver! Any good driver will tell you that your car will follow your hands, but your hands follow your eyes. Driving is not "static."

That is, you are not sitting still. When you are driving on the interstate, **where you are at the moment is not as important as where your eyes are leading you.**

God reminds us in Ezekiel 8 that **you can only face one direction at a time.** The wrong direction comes naturally to us. Going the right direction is only by design. You will never go the right way by accident! Going the right direction is an act of your will as you surrender to God. **You will go the way you decide to go!**

Friend, God knows which way you are heading, even when you do not. Sometimes you might think you are going in the right direction (like the men in Ezekiel's vision), but you are facing the wrong direction. God both knows and cares what direction you are going. Who or what has your attention today? Where are your "eyes" pointing? You cannot serve two masters (Matthew 6:24), and you cannot face two directions at the same time. Where you stand today is important, but where you are focused is where you will end up tomorrow!

◇◇◇

DAY 180

THE PLACE OF WORTH

Ezekiel 11:16b Yet will I be to them as a little sanctuary in the countries where they shall come.

Ah, human nature—how amusing it is! We often find our worth or security in a place or position. Some people find their worth and security when they have a low position in a prestigious place. For instance, a guy may flunk out of his first semester at Harvard, but you can bet your boots Harvard will make it on his résumé! Why is this? He might have had a low position, but he had a low position in a very prestigious place.

Other people get their worth by having a high position in a meaningless place. You might be the president of some obscure association, but you are only the top dog in an unknown or insignificant place! Both examples seem hollow, but they are illustrative of how you and I naturally think. Some Israelites disdained their Jewish brethren who had been taken captive. (Who still occupied Jerusalem.)

I love God's response to the captives who were downtrodden, dismissed, and displaced. He says, *"Although I have scattered them among the countries, yet will I be to them as a little sanctuary in the countries where they shall come."* They weren't in God's Promised Land anymore, but God was still with them. God was "there," no matter where "there" was!

Are you confident in your place or position? Or are you confident in the God who has given you a place in His family? Even Solomon, who built the great Temple to God, did not build a kingdom to himself. He realized that the God of heaven would not be confined by a building or a location on earth. **If God is not your confidence, you have the wrong place and the wrong position!**

There are places that are very special and dear to my life. The Bill Rice Ranch is one of those places. I can think of very important, life-changing decisions I have made on the Ranch, even down to the specific spot. I thank God for those places at the Ranch, but no place should ever replace the God who makes it significant. God alone can provide the security you cannot find in a place or in the merits of your own position. I am not great, and in the grand scheme of things, neither is any position I might attain. **There is no special place unless God is there. If God is there,** *He* **is the One who gives the worth and power we need.**

◇◇◇

PROFITLESS PROPHETS

DAY 181

Ezekiel 13:3 Thus saith the Lord GOD; Woe unto the foolish prophets, that follow their own spirit, and have seen nothing!

It is ironic that in a day when preaching seems to be greatly minimized, the world cannot get enough of authoritative speakers. There is an entire "herd" of leadership conferences, podcasts, and popular events in different venues across the country. When you hear what someone with authority has to say, **how do you know what he says is good?** Is it good merely because it sounds good? Is it good because, in some way, it actually works? **What makes what you hear profitable?**

Ezekiel 13 addresses this age-old question. In Ezekiel's day, there were prophets who spoke from themselves (not from God); and therefore, what they had to say was not useful. They were *profitless* prophets! They might have been well-trained, accomplished, dynamic speakers, but what they spoke was from their *own* spirit and not from God's. God said they *"have seen nothing."* What these false prophets had to say was empty and useless because *they* were the source!

Be sure to judge what you say and what you hear by your source, not merely by the substance. Sometimes things that are actually empty will sound substantial. The way to judge the quality of the substance is to judge the source. If God has said it, I can believe it. The help I try to give other people should be the help that God Himself has given to me.

◇◇◇

DAY 182
COVERING YOUR BASES

Ezekiel 14:4 Every man of the house of Israel that setteth up his idols in his heart, and putteth the stumblingblock of his iniquity before his face, and cometh to the prophet; I the LORD will answer him that cometh according to the multitude of his idols;

Are you aware that it is common practice for big spenders and big businesses to contribute to *both* sides of a major issue? For instance, during a Presidential election, a person or company will give a donation to both the Democratic candidate and the Republican candidate. Talk about covering your bases! While this practice may be frustrating and seem inconsistent to most of us, savvy businessmen make sure that no matter the outcome of the election, they are covered. In other words, **they are playing it safe by playing it broadly.**

Sadly, we tend to do the same thing with God. We trust in God…and our money…and our plans…and our abilities. We figure that if one thing doesn't take care of us, one of the other options is bound to work. I am not saying it is virtuous to be unprepared, bankrupt, and unwise, but Proverbs 21:31 reminds us that *"the horse is prepared against the day of battle: but safety is of the LORD."* When God is all you have, you'll find that He is all you need.

Leaders in Ezekiel's day were trying to cover their bases, so to speak. They made idols in their hearts and with their hands, which became a stumbling block in their relationship with God. Instead of being holy to God, they had separated themselves from God (verse 7). When they didn't get what they needed from a god of stone, they had the gall to come ask the God of heaven for the direction they needed in their lives!

Before you ask God for something, make sure you have a heart that will hear and hands that will obey. How foolish it is to ask God for something when there is an idol in your heart! Any wall I erect between my life and my God is an obstacle between my need and heaven's answer. Thankfully, it doesn't have to be that way! God can answer, wants to answer, and will answer if we ask with a clean heart and an open ear to obey.

◇◇◇

LIVING LIKE ROYALTY

DAY 183

Ezekiel 16:63 That thou mayest remember, and be confounded, and never open thy mouth any more because of thy shame, when I am pacified toward thee for all that thou hast done, saith the Lord GOD.

A story is told of a king who found a young lady on the side of a highway. She was an outcast, uncared for, unpitied, and unclean. The king had compassion on her and took care of her. In fact, she became his bride and queen. Now she had royal apparel instead of the rags she used to wear. She wore costly jewels that reflected the grandeur of her new position. Sadly, as the story goes, she rejected the very person who did not reject her. She was foolish and unfaithful and ended up in a worse position than she was when found by the king.

That is a synopsis of the story God gave to Ezekiel in chapter 16. God is the King of the universe. Like the queen, we are the object of His provision, protection, and care. The queen in the story forgot the one who showed compassion in the first place, and we often do the same. God over and again says, *"Because thou hast not remembered…."*

We are so prone to forget who we were, who God is, and what He has made us now. **We live cheaply when we forget how highly God values us and how much He paid for us.** We live cheaply when we forget how much we owe Him!

If you will live up to the position you have in Jesus Christ, instead of living down to your own will and way, you will live like royalty. Remember the prized position you have and live like one who was purchased at great cost by a great and generous God!

DAY 184 — THE BLAME GAME

Ezekiel 18:2 What mean ye, that ye use this proverb concerning the land of Israel, saying, The fathers have eaten sour grapes, and the children's teeth are set on edge?

There is no doubt that it is human nature to be the first to take the credit and the last to take the blame! Any of us want the praise for a job well done, but none of us want the responsibility before it happens. Rare is the person that will take responsibility; many will lay the blame for some failure or some sin at the feet of other people, including those closest to them. Blaming your parents for your own sins has almost become a common strategy of defendants in the American judicial system. While it is common today, it is not unique to our day. The blame game has been around about as long as people have!

People in Ezekiel's day were so eager to blame somebody else that they blamed their parents. Truthfully, they were blaming God for not being fair. They were not willing to assume responsibility for their own sin; they were willing to assume that God's judgment on them was because of their fathers. They were desperately in need and under God's judgment but unwilling to recognize their need. You can't help a person who doesn't see his need!

The children of Israel had a proverb in that day that said, *"The fathers have eaten sour grapes, and the children's teeth are set on edge."* In other words, the children were suffering the bitterness of life as a result of decisions the fathers had made. What was God's answer to all of this? *"As I live, saith the Lord GOD, ye shall not have occasion any more to use this proverb in Israel."* He doesn't take the shifting of blame lightly!

The fact of the matter is, each of us is accountable to God. While there is no doubt that your parents influenced you, that is never an excuse for you to do wrong. **Your relationship to God is determined by your choices, not your heredity.** Until you come to God and realize that you have a responsibility in choices today, regardless of the influences yesterday, you cannot see a better day. Don't take your turn in the blame game!

◇◇◇

THE HIGH COST OF LOW LIVING

DAY 185

Ezekiel 20:31 For when ye offer your gifts, when ye make your sons to pass through the fire, ye pollute yourselves with all your idols, even unto this day: and shall I be enquired of by you, O house of Israel? As I live, saith the Lord GOD, I will not be enquired of by you.

Suppose your son stole your car, spent your money, and spread lies about you on Facebook. How would you feel? "Brother Wil, I don't even have a son!" Well, shuffle the characters to make it relevant to your life. You get the idea, don't you? What if someone you had taken care of, loved, and guided turned your own resources against you? What if this person lied about you, your character, and dragged your name through the mud? How would you feel? What would you do? How would your relationship be affected?

I am so thankful that **I have a Father in heaven who is perfect, although I am not.** I am not God's child because I do right; I want to do right because I am God's child. If I do wrong, I'm not kicked out of the family, but I set myself up for the judgment of a loving but holy God. There is a high cost for low living!

Israel had experienced the protection of God, enjoyed the provision and direction of God, and inherited a land they did not earn or deserve. They were special because God is extraordinary, and He put His favor on them. Yet they turned on Him! They were barely in God's Promised Land before they started worshipping idols. They were living lower than the position they had as God's own.

You may live like everyone else; but, as God's child, you will never *be* like everyone else. If you belong to God through faith in His Son, it really is a silly question to ask if you should live different than everyone else—**you *are* different!** Whether or not people see that you are different, you are fundamentally different from the world. If you belong to God, but that is not reflected in your life, you are living in hypocrisy.

There is a high cost for low living. When you trusted Christ, God set you apart as one of His own. Your life should reflect that truth every day of your life. What you are in practice you should show in practice. The life you live should tell the truth about your Father in heaven. What is your life saying?

◇◇◇

DAY 186 — A TRUE PERSPECTIVE OF OURSELVES

Ezekiel 27:2 Now, thou son of man, take up a lamentation for Tyrus.

We live in the day of the super-rich, the superstar, and the superhero. While I don't consider myself to be any of those three things, how easily I can be fooled into taking an inflated view of myself! Just one compliment or accomplishment can have me thinking I'm something that I'm not. This is the human tendency, and this pride comes easily for both individuals and their nations. A good perspective helps us have a clear view of ourselves and of life. Our tendency toward pride should be tempered by two things: the brevity of life and the greatness of God.

These two elements are hammered home as you read of the destruction God predicted against Tyre, that wealthy and secure nation. We read of this in Ezekiel 26 and 27. Ezekiel 26:2 says, *"Now, thou son of man, take up a lamentation for Tyrus."* Why would there be a sad song sung for such a wealthy nation? Verse 3 says of Tyre, *"Thou hast said, I am of perfect beauty."*

But there was coming a day of judgment because these people trusted themselves and not God. They saw themselves as great, and didn't see God at all. Verse 27 says, *"All thy company which is in the midst of thee, shall fall into the midst of the seas in the day of thy ruin."* Now what could have helped Tyre have a more modest or accurate perspective of itself? And what could help you and me to do the same?

Realizing the brevity of life helps us maintain a proper perspective of ourselves. We must realize that no matter how much we've accomplished or gained, we're just a drop in the historical bucket. Time happens to all of us. Time should give us a sense of modesty and of our own finite nature.

The second realization that should temper our pride is the greatness of God. The refrain you find throughout the book of Ezekiel is that God declares what He will do in the affairs of men. This is the statement God makes: *"they shall know that I am the LORD."* He is the eternally existing, self-existent One. What He says will come to pass. Those who put their trust in God have access to a power that is unparalleled on earth and undiminished by time.

DAY 187

WHEN WE OWN, WE'RE ON OUR OWN

Ezekiel 29:9 And the land of Egypt shall be desolate and waste; and they shall know that I am the LORD: because he hath said, The river is mine, and I have made it.

How much do you own? Well, it may be that you own very little, or it may be that God has blessed you with quite a stewardship. The more you own, the more you have to worry about.

The empire of Egypt figures prominently into the history of Israel. The very name of Pharaoh conjures up images of wealth, power, and monuments that have survived the ages. So much of that wealth was the result of the Nile River. Egypt had grown accustomed to owning everything. Pharaoh had grown so accustomed to his own power that he thought he owned everything and created the very Nile River that had brought him so much prosperity.

This pride is exactly what set him against God; it is the reason God told Ezekiel to set his face against Egypt and against Pharaoh. Ezekiel was to tell them that God was against them. Pharaoh had said that *his* river was his own and he had made it. Here is a man that had been seduced into thinking that he was the owner of everything in his world. Well, Pharaoh is dead and gone, but God Almighty remains.

The more we grasp as our own, the more worry and anxiety we own. God's words against Egypt remind us that **we are on our own whenever we deny God's ownership.** Peace, security, and perspective all come when we realize that God is the Creator. He owns this world, and we can leave it to Him.

ARE YOU A GOOD SHEPHERD?

DAY 188

Ezekiel 34:10a Thus saith the LORD GOD; Behold, I am against the shepherds; and I will require my flock at their hand.

If I were to ask you to think about the politics of most any major American city, what one word would come to mind? Unfortunately, words like "corruption," "fraud," or "machine" would come to many a mind. Abuse of the public trust for personal gain is pretty much as old as politics itself, and from the beginning of time this has been an affront to God Almighty. To abuse what He has created is to sin against Him.

God spoke through Ezekiel a message of judgment to the leaders of the people of Israel in that day. Why? Because these shepherds fed themselves but not the flock. Ezekiel 34:4 says, *"with force and with cruelty have ye ruled them."*

The shepherd's primary responsibility is to care for the sheep. The only one that should care more for the sheep than a shepherd is the owner of the sheep. God repeatedly shows in this chapter that these sheep belonged to Him. Verse 10 brings all of this into focus when God says, *"Thus saith the LORD GOD, Behold, I am against the shepherds; and I will require my flock at their hand."* God was against the shepherds because they fed themselves; and God would deliver his flock from these shepherds.

These shepherds (leaders) did not own the sheep (people). God did. How do you think an owner of sheep would feel if he employed a shepherd who beat up the sheep, who fed himself instead of the sheep, and who was more concerned about his personal comfort than the safety of the sheep? Well, that is exactly how God feels about those who abuse their stewardship for their own selfish ends.

The fact is that every one of us shepherd or steward something. We're servants of God who have something that God has placed into our hands. How you treat what you have depends on who you think owns it. If you'll remember today that every moment, every dollar, every person, and every project that is in your life belongs to God and has been entrusted to your stewardship, you will have a greater sense of purpose and responsibility than you could ever have going it alone.

◇◇◇

DAY 189
DON'T LET ANGER GROW

Ezekiel 35:11 Therefore, as I live, saith the Lord GOD, I will even do according to thine anger, and according to thine envy which thou hast used out of thy hatred against them; and I will make myself known among them, when I have judged thee.

Ezekiel 35 is the story of how two boys and their families grew, and how their anger grew into national conflict. The brothers were Jacob, who begat Israel, and Esau, who begat Edom. The conflict between these two men began with deceit, carelessness, and the theft of a birthright and blessing; and it grew from there. Long after the memory of the events that began this conflict dimmed in the collective memory of their descendants, the fresh hatred that was still there defined their relationship.

God is responding to Edom's hatred in Ezekiel 35 when the Edomites had taken advantage of the calamity of the Israelites. Verse 11 says, "Therefore, as I live, saith the Lord GOD, I will even do according to thine anger, and according to thine envy which thou hast used out of thy hatred against them; and I will make myself known among them, when I have judged thee." The Edomites were characterized by anger, envy, and hatred. God's response was that He would make Himself known among the people of Israel when He judged the people of Edom.

There are two lessons we can take from God's response. **Any bitterness you allow will crowd the joy right out of your life.** The Bible describes Edom's hatred as perpetual even when God's judgment of wayward Israel had ended. As a result Edom's destruction would be neverending. The Edomites had allowed anger, envy, and hatred to expand in such a way that it crowded joy right out of their lives. It had expanded in time. What had happened centuries before was passed from generation to generation down to these very people. It had also expanded in scope. What had begun as the conflict between two brothers became the conflict of two nations. We must always remember that what we allow today may be the excess of our children tomorrow.

The second lesson we can take from this is the warning that **we should destroy anger before it destroys us.** The Edomites were very proud. They had fortified cities and boasted against God when they gloated against His people. This pride gave life to fresh anger, envy, and hatred every day. It was the very hatred that eventually destroyed not the object of their anger, but the Edomites themselves. Don't let envy and anger crowd out the joy in your life. Seek God's help to destroy anger before it destroys you.

HE'S ALREADY THERE

DAY 190

Ezekiel 38:16 And thou shalt come up against my people of Israel, as a cloud to cover the land; it shall be in the latter days, and I will bring thee against my land, that the heathen may know me, when I shall be sanctified in thee, O Gog, before their eyes.

Which is more likely, someone being proud about his past or his future? You have to be pretty optimistic to be proud of your future! I think you would agree that the world is full of people who love to tell what they have done and what they accomplished years ago. On the flip side, is someone more likely to be afraid of his past or his future? You can reject your past (although you can't change it); but more than likely, the things that concern and worry you are in the future. You are more likely to be proud of your past and fearful of your future.

It is good to remember that your future is already God's history. The God of the universe is not bound by time or deadlines; He is never worried or hurried. God may send clouds your way for your good and His glory. God told Gog, the army attacking Israel, that they were sent so that *"the heathen may know me,* when I shall be sanctified in thee." The army God used was just a pawn in His hand. He declared His purpose: to *"magnify myself, and sanctify myself; and I will be known in the eyes of many nations, and they shall know that I am the LORD."*

I don't know what your future holds, but I know Who holds your future. Your future is already God's history. While you may experience difficulties, troubles, and the storms of life, you can rest assured that God is in control. **Whatever He allows is for your good and for His glory**, so that He is revealed in your life and in the lives of others. You can be confident in your future, no matter what it holds, because God is already there.

◇◇◇

DAY 191
WITH GOD, NOTHING IS UNCERTAIN

Ezekiel 40:4 And the man said unto me, Son of man, behold with thine eyes, and hear with thine ears, and set thine heart upon all that I shall shew thee; for to the intent that I might shew them unto thee art thou brought hither: declare all that thou seest to the house of Israel.

As I read three chapters in Ezekiel, I was reminded of how bored I can be with the tiny details of other people's lives, and how keenly interested I am in my own life. These three chapters are the specifications of a temple that was and is yet to be built. Even though the description of this temple shows that it will be grander than Solomon's Temple, you can still slog through the next three chapters as if you were looking at the blueprints of somebody else's house. It's easy to glaze over and miss the glorious detail of God's plans for the future.

In Ezekiel 1, Ezekiel was taken from the mundane, the present, the shackles of existence as a slave in a far-off land, and was transported in a vision by God Almighty to a high mountain in Israel to witness pictures of the future. God pulled back the curtain and allowed Ezekiel to see, to some extent, as God Himself does.

Why did God go through all the trouble of giving all this great detail to Ezekiel? Well, on one level the answer is given in verse 4 where it says, *"And the man said unto me, Son of man, behold with thine eyes, and hear with thine ears, and set thine heart upon all that I shall shew thee; for to the intent that I might shew them unto thee art thou brought hither: declare all that thou seest to the house of Israel."* God wanted Ezekiel to catch a glimpse of the future so he could give hope to God's bound, discouraged, defeated people.

Ezekiel was very limited. He was a slave living on flat land in the present, and God took him to a high mountain where he could literally see into the future. Now think about your own perspective today. While you may not be literally shackled as a slave, many things can still bind you. Time binds. Perspective binds. The present binds. Life is anything but certain. Too often if we're not afraid; we're just bored. How refreshing to remember that **with God nothing is uncertain—the future, His plans, and His love for you are ever certain!** It's obvious to me that God has our futures planned in glorious detail. Don't be so wrapped up by the present that you fail to be encouraged by the future that God holds in His hands.

◇◇◇

WHO WILL WRITE YOUR STORY?

DAY 192

Daniel 1:8 But Daniel purposed in his heart that he would not defile himself with the portion of the king's meat, nor with the wine which he drank: therefore he requested of the prince of the eunuchs that he might not defile himself.

Winston Churchill once proclaimed, "History will be kind to me, for I intend to write it." While Churchill had no control over the events that came his way, he did have a choice about how he responded to them. He had a choice about what story he would live and what script he would accept. What about your life? Who will write your story?

There is a major difference between genuine faith and helpless fatalism. **Genuine faith acts on the script that God provides and trusts Him for the power to live it.** In the story of Daniel, Daniel was given two competing scripts: the world's (Babylon's) script and God's script. Babylon wanted to teach Daniel their pagan ways, feed Daniel their pagan food, and give Daniel one of their pagan names. In contrast, God gave Daniel favor with others, supernatural knowledge, success, and longevity. Which was the better script?

Every day, you are handed two competing scripts. **You will either be pressed into the mold the world demands, or you will live the life that God made you to live.** Daniel's choice was purposeful. *"But Daniel purposed in his heart…."* Daniel's mission statement was clear: *"that he would not defile himself."* Daniel's purpose was tested: *"Prove thy servants."*

Like Daniel, you can choose which script you will accept. Is your choice purposeful? Is your purpose clear? You can be sure that your choice to live for God will be tested by the world, but don't let that stop you from obeying what God has given you to do!

We become victims rather than victors when we accept the world's script rather than acting on God's. When it comes to your storyline, what script will you follow? **Who will write your story?**

GIVE CREDIT WHERE CREDIT IS DUE

DAY 193

Daniel 5:3-4 Then they brought the golden vessels that were taken out of the temple of the house of God which was at Jerusalem; and the king, and his princes, his wives, and his concubines, drank in them. They drank wine, and praised the gods of gold, and of silver, of brass, or iron, of wood, and of stone.

Daniel 5 records the crowning act of arrogance of an empire represented by King Belshazzar. God had brought him to power, and God was about to bring him down. Yet he was so smug that he threw an epic party for a thousand of his lords. The Bible records that they all drank wine out of the golden vessels that had been taken out of the Temple of God in Jerusalem. Verse 4 says, *"They drank wine, and praised the gods of gold, and of silver, of brass, of iron, of wood, and of stone."*

This act of arrogance was just hours before the utter destruction of this empire. Their arrogance had not only disregarded God, it had also misattributed God's power to other people and other gods. It wasn't enough that Belshazzar didn't recognize God's role in his empire. It was not enough that he claimed God's vessels from the Temple as his own. It was not enough that he filled God's vessels with his wine. No, he seemed to be praising the vessels themselves.

At the very best this was naïve of him, as arrogance always is. If you look at Daniel 1, the Bible explicitly tells us that it is God Who put the vessels of gold from the house of God into the hand of the Babylonians.

Now, there's a lesson to be learned here. **Your time is very nearly gone when you refuse to give credit where it is due.** This act of arrogance immediately preceded the final gasp of the Babylonian empire. Later in Daniel 5, we learn that Belshazzar knew how God had humbled Nebuchadnezzar when he had been lifted up in pride. Belshazzar saw what pride had done to Nebuchadnezzar but still lifted himself up against the God Who held his very breath.

It's fascinating that on this night in which this king attributed God's power to his own gods and filled God's vessels with his own wine, he asked for help from another vessel of God, Daniel. Three times in these verses the Babylonians recognized the Spirit of the Holy God that was in Daniel; but even this they misattributed to their gods. Daniel had knowledge, wisdom, and power that all came from the Spirit that was within him. In both cases, the king glorified the vessel (the vessels of gold and the vessel Daniel) and ignored God.

We need to be very mindful that whether we seem to be made of gold, silver, or some other element, **our value does not come from what we are made of. It comes from Who made us.** We have value to others when we allow ourselves to be filled with God Himself, to be His vessel, and to give Him the credit for all that is good in life.

◇◇◇

DAY 194
POWER IS FROM SUBMISSION, NOT REBELLION

Daniel 6:10 Now when Daniel knew that the writing was signed, he went into his house; and his windows being open in his chamber toward Jerusalem, he kneeled upon his knees three times a day, and prayed and gave thanks before his God, as he did aforetime.

Some readers are cheaters. Maybe you are one of those readers who doesn't possess the patience to read the book all the way through before you find out how the plot plays out. Well, if that's your weakness, I want to invite you to indulge today by looking at the very last verse of Daniel 6, the story of Daniel in the lions' den.

The Bible says in Daniel 6:28, *"So this Daniel prospered."* Now there's no reason at all that Daniel should have prospered, humanly speaking, given his Jewish race, his slave status, and the fact that he was going against the grain in a totalitarian government and wicked culture. Now this doesn't mean that Daniel was a rebel. He certainly was not.

The other presidents and princes under his authority resented him and sought to get him in trouble. They knew he would not be a rebel and willfully break the law, so they made a law he would have to break because he was living in submission to God. The law was that no one was to pray to anyone or any god save the king. The king naively signed this piece of Medo-Persian legislation which could not be changed. Then they waited to spring the trap.

I was fascinated to read in verse 10, *"Now when Daniel knew that the writing was signed...."* Note those words "when Daniel knew." Daniel prayed precisely when he knew this action was taking him into dangerous waters. This was an open act, not of rebellion, but of submission. The king had no choice but to throw Daniel into prison; but he did so with a broken heart because he loved Daniel.

Our power comes from submission to God, not from rebellion towards men. This submission dictates our stance towards men's laws. Daniel was driven by obedience, not rebellion. It's obvious that these princes, the actual rebels of the story, knew that Daniel was a man of submission. The king himself knew that Daniel served God; so in some serve the king understood that Daniel was acting under submission to God, not out of rebellion towards him. Truly, Daniel was more submissive to the king in his heart than the princes and presidents were who had concocted this law.

You are going to have the courage to stand when the chips are down if you habitually live in submission to God. There's a big difference between practicing obedience towards men when it is safe and living in obedience to God when it is not. Our power comes from submission. May that submission give us power before men and before God.

DAY 195
A TIMELESS PERSPECTIVE COMES FROM AN ETERNAL GOD

Daniel 7:14 And there was given him dominion, and glory, and a kingdom, that all people, nations, and languages should serve him; his dominion is an everlasting dominion, which shall not pass away, and his kingdom that which shall not be destroyed.

Nearly all of us have this fascination with "the next big thing." Some people are interested in the next big technology. For others, "the next big thing" would be a social event because they are social butterflies who want to keep up with who is doing what. Others want to make sure that what they're wearing is up to date. They want to be the first person to have the new shoes. Many of us are interested in the newest current events.

So what if someone could give you a high-definition, large screen, heads-up display of "the next big thing?" What if God could give you a birds-eye view of the next four world powers? This is exactly what God did for Daniel in a vision. Daniel 7:4 describes the Babylonian empire. Then the subsequent verses describe the Medo-Persian, the Greek, and the Roman Empires.

In verse 14, God's Son, the Messiah, came in a vision to Daniel. The Bible says of the Messiah, *"And there was given him dominion, and glory, and the kingdom, that all people, nations, and languages, should serve him: his dominion is an everlasting dominion, which shall not pass away, and his kingdom that which shall not be destroyed."* The kingdom of God's Son is unlike that of any other. It includes all people of every nation and language, and they all will serve Him. This everlasting kingdom will not pass away and will not be destroyed like those of the Babylonians, Medo-Persians, and the Romans.

If we are forever consumed by "the next big thing," then ironically, our perspective will be anything but big. It will be narrow, confined by time and space and by the limited parts of history that we can actually perceive. Above it all is the great God, the Ancient of Days. He is not merely ancient: He is timeless. God

Almighty straddles the present with a foot in the ancient days and a foot in the future. **A timeless perspective comes from a focus on a great and eternal God.**

I don't know what the next "big thing" will be for you, but let's remember that God is greater than time, unlimited by days, and unlimited by the power, understanding, and resources of man.

◇◇◇

PRAYER: BARTERING OR BEGGING?

DAY 196

Daniel 9:18 O my God, incline thine ear, and hear; open thine eyes, and behold our desolations, and the city which is called by thy name: for we do not present our supplications before thee for our righteousnesses, but for thy great mercies.

Do you know the difference between bartering and begging? Bartering is when I have something of value and I see something of value that someone else has and we agree to trade. When I was a kid, I didn't have money, but I did have Matchbox cars. So, oftentimes, my friends and I would trade Matchbox cars.

Now it's a general maxim of negotiation that you can only negotiate from a point of strength. You cannot ask for what someone else has unless you have something of equal value with which to trade. Begging is different. Begging happens when you have nothing and you need something.

The distinction between begging and bartering is brought into clear focus by the prayer of Daniel to God in Daniel 9. Notice what Daniel claimed to own and what he knew God owned. In verse 7 he says, *"O Lord, righteousness belongeth unto thee."* It didn't belong to Daniel, nor his people, but to God. Verse 8 says, *"O Lord, to us belongeth confusion of face."* Now Israel owned something, but it wasn't righteousness. It was wickedness and confusion. Verse 9 says, *"To the Lord our God belong mercies and forgivenesses, though we have rebelled against him."*

Now if you continue to read Daniel's prayer, you'll see a wonderful picture of the very nature of prayer. Daniel 9:18 says, *"O my God, incline thine ear, and hear; open thine eyes, and behold our desolations, and the city which is called by thy name: for we do not present our supplications before thee for our righteousnesses, but for thy great mercies."* **Prayer is coming to terms with what truly belongs to us and what truly belongs to God.** Until you come to a point when you realize your own bankruptcy, your own need, and in contrast to that, God's righteousness, mercy, and forgiveness, you're not going to come to God in prayer.

◇◇◇

DAY 197
THE ONE WHO HOLDS THE FUTURE

Daniel 12:8-9 And I heard, but I understood not: then said I, O my Lord, what shall be the end of these things? And he said, Go thy way, Daniel: for the words are closed up and sealed till the time of the end.

God's vision of the future is just as certain as if it were already a memory. But if you do not fully comprehend what God says about the future, you are not alone. I am not so sure that Daniel knew completely either! While I think Daniel understood the broad strokes of what God revealed to him about the future, I am sure he did not completely understand the specific timing of everything.

The Bible is meant to be understood. The Bible is meant to be carefully studied. We would all do well to study the Bible, including the prophesy portions of Scripture. However, **our peace for the future should not come in understanding every sign; it should come in knowing the One that gave the signs!**

You may know that I travel quite a bit with my family. Suppose we are driving to California. What do you think my eight-year-old asks me an hour into the trip? Like your kid probably does, he says, "Dad, are we there yet? How far to California?"

Now that is a question that has an informed answer, and that is one question that I can actually answer! I can answer Weston's question by either pointing to a road sign that says, "California—2,314 miles," or I can encourage him to sit tight because the truck he is riding in is heading to California. Weston doesn't comprehend miles or what is involved in the trip; but he doesn't have to understand every road sign if he knows his father. If his dad makes it to California, and he is with his dad, Weston will make it too!

Friend, you don't have to understand every sign if you know the Lord who gave them. Daniel expressed some anxiety about not understanding, but he expressed his anxiety to the very God who was already in the future! My confidence does not come because I know every minute detail of the future; it comes because **I can talk real-time, at any time, to the One who holds the future!**

SIN: A DEPARTURE FROM GOD

DAY 198

Hosea 1:2 The beginning of the word of the LORD by Hosea. And the LORD said to Hosea, Go, take unto thee a wife of whoredoms and children of whoredoms: for the land hath committed great whoredom, departing from the LORD.

One day John came home to an empty house. His wife was gone. The letter she left indicated that she no longer loved him and that she had left him for another. In another town, Jennifer's husband said that he no longer loved her. It's not that there's another person, and there are no hard feelings. It's just that he no longer wants to be married.

We're living in a day where we have made nothing of marriage. In Hosea 1, God powerfully uses marriage to illustrate His relationship to His people. He gave the prophet Hosea a unique task, to marry a woman who Hosea knew would be unfaithful to him. Why would God demand such a thing? Why would

Hosea do such a thing? And why would God record it all in full color in the book of Hosea?

It's painful to think about how terrible this episode would have been for Hosea and his unfaithful wife; but if you're just feeling sorry for Hosea, you're missing the point. The unfaithfulness of Hosea's wife pictures our unfaithfulness to God when we sin. God takes sin personally because He loves us personally. **Sin is departure, not from a code or set of laws, but from a person, God.** I can no more say to God Almighty, "Hey I'm going to sin and do my own thing, but don't take it personally. I still love you; we can still be friends," than I can honestly say that I'm going to leave my wife with no hard feelings and that we're still going to be friends. That's not honest, and that's not the way life or love works.

Sin, as a departure from God, is seen in three ways:

Unfaithful- It's showing unfaithfulness to the God Who hates sin so much but loves me so much that He sent his only Son to take my punishment and place.

Ungrateful- In chapter 2, Hosea says that his wife doesn't comprehend the things he provided for her as her provider and protector. Now the parallel is clear. When I sin against God, I'm ungrateful. God provides, protects, and directs. For me to dismiss these tokens of His love for me is not just shallow or unthinking; it's ungrateful.

Unfruitful- Now this description is not completely true because I will always reap what I sow, whether good or bad. But sin is unfruitful in the sense that nothing good can come of it—for me, for those I love, and for God.

We're living in a day when people are constantly seeking something better and more fulfilling than the marriage they have, only to find that it leads to heartbreak. But that's only a picture of the kind of heartache, lack of fulfillment, and fruitlessness, that comes when I sin against the God Who loves me and cares for me.

WHAT YOU DON'T KNOW CAN HURT YOU

DAY 199

Hosea 4:6 My people are destroyed for lack of knowledge: because thou hast rejected knowledge, I will also reject thee, that thou shalt be no priest to me: seeing thou hast forgotten the law of thy God, I will also forget thy children.

My family and I do a lot of traveling in a year. Most of the time, we take our "home" with us by pulling a fifth-wheel trailer behind us. Recently, while staying in our trailer, I woke up in the middle of the night and needed some fresh air. Sleepily, I reached up and slid open the window by my bed and then drifted off to sleep.

The next morning, I noticed a bloody cut on my index finger that was causing noticeable pain. I had no idea what I had done! Then I vaguely remembered messing with that window in the middle of the night. Evidently, I had a very sharp problem on that window!

No one enjoys pain. I don't like toothaches, body aches, or broken bones because I don't like pain. **But honestly, pain is not always the problem; pain is designed to point you to the problem.** The worst place to be is in a dreadful condition but unaware of it because there is no warning mechanism. The worst problems are the ones that don't look like problems!

The people God addressed in Hosea were looking to idols to answer their problems. The idols weren't the answer; the idols were the *problem*! The pain that Israel suffered was not the problem; the pain was sent and allowed by God to point people to their need and God's answer. God said of them, *"My people are destroyed for lack of knowledge: because thou hast rejected knowledge, I will also reject thee…."* What they did not know did, in fact, hurt them!

The worst thing that can happen to you is to think you are fine, while out of fellowship with God Almighty. God allowed pain in Hosea's day in order for His people to realize their problem. You cannot find contentment while forgetting the One who gives it. God designed you, He made you, and He provides for you. He provides both provision and pain so that you will look to Him.

◇◇◇

DAY 200 — LET GOD LEAD

Hosea 8:3 Israel hath cast off the thing that is good: the enemy shall pursue him.

When I was in the Philippines a few years ago, I was driven to various places where I couldn't find my way back if my life depended on it. I was on an island that I'd never been before, in a city that I'd never visited before, and I was on streets that seemed to me to be chaotic. Yet I had a driver who knew where he was going, so I was ok.

There are two kinds of people today. There are those who are led and those who are lost. Actually, all of us are following somebody; but what I mean is that all of us either allow God to lead us, or we are lost to our own way. Those who refuse God's leadership are lost to their way.

Such was the case with God's own people Israel. When Hosea says in Hosea 8:3 that Israel had cast off the thing that was good, he means they had cast off God, His good guidance, His good leadership, and His good provision. Israel had cast God off, and now their own idolatry, wickedness, and rebellion would cast them off. Their own ways would surround them, beset them, and destroy them. They were lost. Oh, they had leadership, but it was the wrong kind. Oh, they had religion, but it was one of their own making. They were not led by God. They were lost to their own way.

Now if this calamity could befall an entire nation of people, don't you think it could befall one person? Where are you with God? Are you being led by Him? Or are you lost? Perhaps you are considering some major decision or even a small decision that gnaws at your mind. Few things are worse than being lost to our own ways.

Increasingly, to follow God's leadership is a frightening thing because it is so counter to the culture and what we read, hear, and watch every day. But God's Word works and God's way works. **Let God lead.** Seek Him. Be willing to follow Him, and I can promise you He will show you the way.

◇◇◇

TO WHOM ARE YOU DEDICATED?

DAY 201

Hosea 9:10 I found Israel like grapes in the wilderness; I saw your fathers as the firstripe in the fig tree at her first time: but they went to Baalpeor, and separated themselves unto that shame; and their abominations were according as they loved.

August 27, 1994, was one of the greatest days of my life. It was on that day that my wife and I dedicated ourselves one to another in marriage. Marriage is one of the first creations of God. It is one of the primary foundations of civil society. It is one of the greatest delights of life. And it is the defining relationship of a family. There is something terribly wrong when someone sees marriage as restrictive or as a drudgery, and not as a delight.

Drudgery is exactly what Israel's relationship with God Almighty had become. She had wearied of Him and turned from Him. She wanted a human king instead of God's leadership. She'd gone to pagan gods that had been made by man instead of looking to the God Who made man. She put her dependence on world powers instead of the Power Who made the world. The Bible compares Israel to an unfaithful spouse.

Many people live dedicated and separated lives. A man in the NFL is dedicated to his sport and to disciplining his body to become strong and fast. A lady in grad school is dedicated to learning and getting a degree. A concert pianist is defined by music. In every case, this **dedication means separation.** For an NFL player, there are some things he cannot eat if he's going to be the best. There are some things that a grad student doesn't have time to read. And a concert pianist doesn't have time to listen to just any kind of music. To some degree, we all live dedicated lives. So, to whom are you dedicated, and from whom are you separated?

Your dedication determines your separation. The thing or person to whom you're dedicated determines the thing or person from whom you are separated. My life ought to be dictated by my love for God, but if it is, then it will also be defined, to some extent by my hatred of that which is in opposition to my relationship with God. That's why the Bible tells us not to love the world. The world system is in opposition to God.

Make no mistake, everyone lives a separated life. **To whom are you dedicated, and from whom should you be separated?**

◇◇◇

DAY 202
THE HEART OF THE MATTER

Joel 2:12 Therefore also now, saith the LORD, turn ye even to me with all your heart, and with fasting, and with weeping, and with mourning.

A few years ago, my wife and I were ministering at a church in Connecticut. One day that week, I drove a couple hours away to Massachusetts to preach in a school chapel. Coming back from this little trip, I arrived in the town where we were staying. I came into the center of town and wasn't quite sure which direction to turn. I'm ashamed to say this, but I ended up driving in circles for about half an hour before finding my way "home"! Each right or left turn I took had buildings that looked familiar, but the way didn't lead to my borrowed home. Finally, I drove back to the interstate and got off on the correct exit to carefully retrace my directions to the home in which we were staying.

Now, would I have been content just to be in the town where we were staying or just to be around familiar buildings? No—I was only content when I arrived "home" where my wife was.

Here in Joel, the children of Israel remained away from God. They had turned from Him long ago and were reaping the awful fruit of rebellion. The locusts had come and destroyed their food supply, casting the land into a famine. It was at this time that the prophet Joel carried God's message to His people: *"Turn ye even to me with all your heart…."*

They were experiencing great hardship, but the real problem was not the locusts or the lack of food—**the root problem was their relationship to God.** The same is true today. Many hardships that people experience today have their root in their relationship to God. God does not change! He desires for His people to walk in fellowship with Him; and when they choose to turn away, they will experience hardship and emptiness.

How are things in your life? If they are not right, turn back to God. Notice what the Lord says in the first part of verse 13: *"And rend your heart, and not your garments…."* It is not good enough to just turn <u>from</u> your wicked ways. The Israelites were good at making a big show of tearing their garments and putting ashes on their heads in outward "repentance." **God was not interested in a show—He wanted their hearts.**

Outward reform, both being good and doing good works, will not set things right in your life. Only when you return in your heart to fellowship with God will you find the missing peace.

One time, my wife fixed a delicious spaghetti supper for our family. Unfortunately, my children became suspicious because the sauce contained onions, a vegetable they try to avoid at all costs. I told them they were to quietly eat their spaghetti without another complaining word about the onions. I was not so concerned about the food actually going down into their stomachs and giving them nutrition (though this outward action was good for them too); I was more concerned about their attitude and their heart in the matter. I was serious about their learning how to obey and to be thankful for what is set before them.

So it is with God. **Yes, outward reform is important and good, but it is nothing without inward repentance toward God.** He wants our hearts! And verse 13 gives us a compelling reason to turn back to God: *"For he is gracious and merciful, slow to anger, and of great kindness...."* Aren't you glad that our God is this way?

When I was lost, being in a familiar place wasn't good enough for me—I wanted to be home. When you are away from God, just turning from your sin won't do—you need to go home! **Give God your heart today and find that He is what you have been missing.**

◇◇◇

DAY 203
GOD'S CORRECTION

Amos 3:2 You only have I known of all the families of the earth: therefore I will punish you for all your iniquities.

My wife and I were behind a very unhappy mother with two bad kids in a Walmart in California the other day. The little girl was kicking and screaming while Mom tried to drag her along. (This very scene has probably been played out in Walmarts across America!) The boy was in a cart in front of us; and when my wife made eye contact, he wrinkled his snotty little nose and stuck out his tongue! Now, it is true that I was happy neither one of those kids were mine, but I didn't do anything about it. He is not my son, and she is not my daughter!

On the other hand, upon being picked up from the airport, I noticed my kids were doing something that I did not want them to do. Sena and I had been away and correcting my kids was not at the top of my list, but I had a responsibility to correct what was wrong because they are my kids!

God, as our Father, corrects us, not because He hates us but precisely *because* we belong to Him! God reminded the children of Israel in Amos 3 that they belonged to Him; and because of this, He was chastising them. A brat in Walmart who disrespects me can never compare to Wilson, Weston, or Lauren Rice! Because I love my kids, I must correct, teach, and train them.

It would be a mistake for one of my kids to think, "If anyone will give me a pass about doing wrong, my dad will." Actually, if anyone will not give him a pass, his dad will not! Likewise, God could not excuse Israel's sin, and He will not excuse our sin today.

The recognition that God chastens His own should protect you from a couple of misconceptions. The **first misconception** is that since God will judge this world, He doesn't care what you do. God does care, precisely because you are His child! Don't be deceived!

The **second misconception** is that since things are not going well, God must not love you. Sometimes difficult times come precisely *because* you are doing what's right. Don't be discouraged!

As our Heavenly Father, God provides protection, provision, direction…*and* correction. **Do not be deceived and do not be discouraged—God loves and chastens His own!**

"For whom the LORD loveth he correcteth; even as a father the son in whom he delighteth." (Proverbs 3:12)

GOD'S PERSISTENCE

DAY 204

Amos 4:8 So two or three cities wandered unto one city, to drink water; but they were not satisfied: yet have ye not returned unto me, saith the LORD.

One day when Weston was just a little guy, I reminded him to thank his mother for the lunch he had enjoyed, but he refused to do it. Now, that is a silly thing to get in trouble for, and he sure did get in trouble for it! The "thank you" was not the point; the refusing was. The ordeal was a little uncomfortable for me as a parent, but there was a definite change after that. Weston's will had to be broken.

Having your stubborn will broken is something that is necessary but not fun. We learn in Amos that **God matches our stubbornness with His persistence.** When we are stubborn in our will, He is persistent in His love to pursue us. In Amos 4, God had allowed His people to go hungry and thirsty because He loved them, while they stubbornly refused to return to Him. The problem was not the idolatry, the immorality, or the oppression; the problem was that they would not return to God.

God was persistent to pursue His people, and He will be persistent to pursue you today. **God is not about "beating you over the head"; He is about having you turn to Him!** Have you ever known someone that was running from the Lord? All the "coincidences" they notice are really God's persistence!

Your will is the battlefield. Your problem really is not gossip, thanklessness, or a bad attitude; your problem is your *will*. **And the solution to your problem is turning back to God!** If I, as a sinful father, would be persistent with Weston about being thankful, don't think that God will do less with His children. Remember that God will match our stubbornness with His persistence.

◇◇◇

DAY 205
FROM BAD TO WORSE

Amos 5:18-19 "Woe unto you that desire the day of the LORD! to what end is it for you? the day of the LORD is darkness, and not light. As if a man did flee from a lion, and a bear met him; or went into the house, and leaned his hand on the wall, and a serpent bit him."

Do you know the meaning of the phrase "from bad to worse"? Do some examples come to mind? Bad is walking on thin ice; worse is falling in! The Bible gives us another example of bad to worse in Amos 5—while running from a lion, a man runs into a bear! And when he finally gets to the safety of his house, there is a snake ready to strike him! I don't know what you think of when you hear "from bad to worse," but that scenario is pretty rough!

What is even worse than the story above—and what this passage is getting at—is thinking, "I can't wait until the Lord judges this wicked world," when a little bit of the wicked world is in your heart. God says, *"Woe unto you that desire the day of the LORD! to what end is it for you? the day of the LORD is darkness, and not light."* Bad to worse is thinking that you want God's judgment on this world when you are not right with Him yourself!

The people of Amos' day were concerned with going to a place to worship the Lord, while He was (and is) more concerned about the condition of their hearts. These folks thought that they could pacify God with the religious things they were doing, yet their heart was not right with Him. In verses 21-23 God says of their religious deeds, *"I hate, I despise your feast days, and I will not smell… I will not accept them… Take thou away from me the noise of thy songs; for I will not hear.…"*

They thought that worship was something you have in your hand or a place to which your feet go, but **worship is in the heart.** They were sacrificing, going to Bethel, and even making beautiful music, but they were also taking bribes and oppressing the poor (verse 12). They were not right with God because their *hearts* were not right. All the good deeds they did and the sacrifices they made could never justify their wicked hearts.

God is not impressed by what you do, if your heart is not right. **You are not right with God if you are not right with other people.** No amount of good deeds can make up for a wicked heart toward others. Wishing for God's judgment on the wicked when your heart is not right is going from bad to worse!

We should live a life that does not need to fear the coming judgment of the *"day of the LORD."* The way to live that kind of life is to have a heart that is right with God and right with others. That is the kind of heart that God will respond to and provide for.

DAY 206

ARE YOU PLUMB?

Amos 7:7 Thus he shewed me: and, behold, the LORD stood upon a wall made by a plumbline, with a plumbline in his hand.

Everyone today is big on "balance." It seems everyone thinks he is, or wants to be, balanced. Balancing work and home life. Balancing your budget. Balancing your diet. Balancing your social life. You could probably write a book or two about those topics, don't you think? Everyone is so enamored with balance, but exactly what is it?

A literal balance is valuable because it centers on an absolute standard—gravity. Most people, in their quest to achieve balance, often look at the wrong standard. Balance is not defined by cutting a straight line between two extremes. Balance is aligning with the truth. **Balance is not determined by where you find yourself in comparison to others; balance is determined by where you are in comparison to God.**

The people of Amos' day thought they were fine. They were religious and prosperous, financially. By the standards of religion and money, they seemed to be doing well. However, in relation to the law of God, they were not "plumb."

According to the dictionary, a plumbline is "a line from which a weight is suspended to determine verticality or depth." If something is "plumb," it is straight and aligned with earth's center of gravity. In other words, it is balanced. In construction, you use a plumbline to check your work and make sure it is straight. The absolute standard is based on God's law of gravity, not another man-made building or landmark or the surrounding terrain. You can't get straight and plumb unless you align with the standard!

In life, **God's Word functions as our plumbline.** You are not balanced just because you strike a middle position between two groups of people. What society considers balanced is always shifting and changing because it is based on extremes instead of an absolute standard. To be balanced, do not look at two extremes and then take the middle ground. No, align yourself with God's truth, and you'll find that you are balanced because you are plumb with His standard!

MERCY AND HUMILITY

DAY 207

Jonah 3:5-6 So the people of Nineveh believed God, and proclaimed a fast, and put on sackcloth, from the greatest of them even to the least of them. For word came unto the king of Nineveh, and he arose from his throne, and he laid his robe from him, and covered him with sackcloth, and sat in ashes.

Who is the worst person you have ever known? Does he or she deserve God's mercy? If so, why? If not, why not? Better yet, why do *you* deserve God's mercy? If you don't think you need it, you are in trouble! If you think you deserve it, you are in big trouble! By very definition, mercy is undeserved. Mercy is not getting what you do deserve.

If you had known the king of Nineveh, you would've probably considered him a very evil and powerful person. He would have been the man least deserving of God's mercy!

Guess what? He received God's mercy! Why? Because **one does not receive God's mercy by being innocent; one receives God's mercy by being humble.** The Bible tells us that *"God resisteth the proud, but giveth grace unto the humble."* (James 4:6) In Jonah 3, the wicked king of Nineveh responded to God's message through the prophet Jonah. He didn't see himself as good; he saw himself as needing mercy! The repentance started with the king and spread *"from the greatest of them even to the least of them."*

It amazes me that some people can be so proud of being good. Have you ever met someone who was proud of how honest he was? You won't impress God by how good or deserving you are. Ironically, the one thing you can't be proud of is being humble! **Humility is submissively seeing yourself as God sees you.** You will never come to the start of God's mercy until you come to the end of yourself.

You don't receive God's mercy by being innocent. You don't receive God's mercy by being good. You receive God's mercy by being humble.

◇◇◇

DAY 208

WHAT YOUR EMOTIONS SAY

Jonah 4:4 Then said the LORD, Doest thou well to be angry?

What is important to you? Think about what you hear, see, and feel! A mother waking to answer a newborn's cry (while dad is fast asleep), a shopper buying a certain product (because of a familiar label), or an athlete crying tears of joy after winning a championship trophy all reveal that emotions are closely intertwined with what is important to a person.

The things that elicit your emotions expose your priorities. Consider Jonah. He was a man full of emotion, and you see a wide range of emotions in Jonah 4. Let's consider each emotion and apply it to our lives today.

First of all, **what makes you glad?** Jonah was *"exceeding glad"* about a gourd that provided shade for him (verses 5-6). In essence, he was happy about comfort. Comfort was his goal. What are your goals? Do your goals revolve around people or things? **Only people last; only people matter.** Thank the Lord that someone made *you* a priority before you were saved!

Second of all, **what makes you angry?** Jonah was angry *"even unto death"* because the Lord took away the gourd (verse 9). Ironically, Jonah was angry because God was slow to anger and showed mercy to others. He was upset that Nineveh received God's mercy when, in fact, *he* had also received the same mercy in the belly of a big fish! What does it take to make you angry?

Third of all, **what makes you take pity?** Jonah had pity on a silly gourd that meant nothing. Today, there are several charitable causes that seek your pity (world hunger, malaria, etc.), but

I wonder how much money we are spending on the gospel? How much time are we spending spreading the gospel? Jonah had pity on a plant that helped no one except himself. What is important to you?

Jonah was a man driven by emotion. He felt happiness, anger, and pity all in the same chapter. The lesson to remember is that what elicits strong emotion exposes your priorities. What do your emotions say about your priorities?

DEPENDENCE OR PRESUMPTION?

DAY 209

Micah 3:11 The heads thereof judge for reward, and the priests thereof teach for hire, and the prophets thereof divine for money: yet will they lean upon the LORD, and say, Is not the LORD among us? None evil can come upon us.

I hope you know that you can trust God with your life. Can you trust God with your future? Yes. Can you trust God with your home? Yes. Can you trust God with your health, your finances, and your friends? Yes. Can you trust God to cover for you when you do what's wrong and rebel against Him? Uh, no. The Bible tells us that when we confess our sin, God's forgiveness covers it; but to live in rebellion and think that things will turn out well is not trusting God. Quite to the contrary, it is presumption. It's being presumptuous against God. There's a profound difference.

Micah highlights this difference in Micah 3:9 when he has strong words for the leaders of God's people. He says, *"Hear this, I pray you, ye heads of the house of Jacob, and princes of the house of Israel, that abhor judgment, and pervert all equity."* He's talking to the leaders who were rebelling against God's authority and thinking that it was ok.

Verse 11 shows that the judges didn't care about justice but took bribes instead. The priests had no passion for God's truth but only had a desire for money. The prophets didn't want to hear from God but desired to be paid by men. Despite all this sin, their attitude was that they were God's people and God wouldn't allow anything bad to happen to them. That's not dependence; that's presumption.

The key difference between trusting in God and presuming upon God is our obedience to Him. In verse 8, Micah was confident that he had power, discernment, judgment, and might because these things belonged to God, and he was acting on God's orders. Let no one think that serving God in some way enables one to do whatever he wishes without consequence. That's not trusting God. That's rebellion, presuming against God. **Trusting God is living in obedience to Him and depending upon Him for the result that will come.**

DAY 210
BEWARE OF DRIFTING

Micah 6:8 He hath shewed thee, O man, what is good; and what doth the LORD require of thee, but to do justly, and to love mercy, and to walk humbly with thy God?

Can you think of anyone that you liked two years ago that, quite honestly, you don't care for now? That's a terrible but fair question. Aren't there people you have drifted from over time? You may not even realize that you're not as close as you once were. Like a bottle on the tide, humans are prone to drifting. We're going one direction or the other at all times. It's important to know which direction we're headed and to go that way deliberately, not by drifting.

Drift often defines our relationship with God as well. If I were to ask you where you are with God right now, it may not be something you've consciously thought about recently. You may have drifted and not even realized it. Some people even become sour or cranky with God, and they don't really realize that it's happening because it happens over time. While we are often oblivious, God is always mindful of where He is with us because this relationship is important to Him. We are His creation, and this matters to Him.

There are two questions that may help reveal drift in our lives. The first is to ask, "What has God done?" The second is to ask, "What does God want?" These are questions that God invited His wayward people Israel to consider in Micah 6:3. He says, *"O my people, what have I done unto thee? and wherein have I wearied thee? testify against me."* How astounding that God asks these people to tell Him their reason for souring on Him!

Throughout Micah 6, God reminds Israel of all the things He had done for them that they had forgotten. God gave rescue, leadership, blessing, manna, and mercy. God had also saved them from destruction many times. It wasn't that God hadn't done anything for Israel. It's that Israel didn't remember or even notice.

The second question is, "What does God want?" God answers that question in Micah 6:8 where the Bible says, *"He hath shewed thee, O man, what is good; and what doth the LORD require of thee, but to do justly, and to love mercy, and to walk humbly with thy God?"* Do you know what God wants from you? You! God doesn't want what you have. He gave it to you. God wants you.

Are you adrift? It might be a good time to ask yourself, "What has God done for me?" and "What does God want from me?" Ask God to help you answer those questions. It will make a difference in the way you see God, the way you respond to Him, and the way you perceive the world today.

DAY 211

DON'T LET GOD'S MERCY BE A JUDGMENT ON YOU!

Nahum 1:15 Behold upon the mountains the feet of him that bringeth good tidings, that publisheth peace! O Judah, keep thy solemn feasts, perform thy vows: for the wicked shall no more pass through thee; he is utterly cut off.

Ahh, how refreshing to come to the book of Nahum! When you read through the prophets, there are many pictures of God's judgment upon evil. Then you come to the book of Nahum. We know little about the man, but we know much about God's message because He gave it to this man whose name means *comfort* or *consolation*. How do you suppose a book named *Comfort* must read. Well, the first verse says, "The burden...." Well, we're two words into the book and already it doesn't look very good. How can there be any burden in a book titled *Comfort*?

Verse 2 has an even worse beginning. It says, *"God is jealous, and the LORD revengeth...and he reserveth wrath for his enemies."* Like the book of Jonah, this book addresses the people of Nineveh, the precursor and later capital of the mighty, brutal, and ruthless Assyrian Empire. God was going to judge Nineveh one hundred years before Nahum when He sent another prophet, Jonah, to proclaim judgment. The judgment had been postponed because the people repented. They turned to God, and God gave them mercy.

We don't receive mercy by being good or innocent. We receive mercy by being humble and honestly agreeing with God about ourselves. One hundred years after God gave that mercy to Nineveh, they were right back where they had been before. Once again, Nineveh became wicked, godless, oppressing, and cruel to the people they subjugated. But God will not let the wicked go without justice. Nahum 1:8 says, *"He will make an utter end... and darkness shall pursue his enemies."*

So this book called *Consolation* is just as much about judgment as it is about mercy. There's no such thing as mercy if there is no such thing as judgment. Mercy is God's withholding of His righteous judgment. If the book of Nahum is largely about God's judgment on Assyria and Nineveh, then why would it be called *Consolation*? Who received consolation in this book? Did anyone? Yes! God's judgment on these remorseless Ninevites was God's mercy on the people they were afflicting. Nineveh's judgment was Israel's consolation. In that way, God's judgment was consolation.

You don't ever want to put yourself in a position where God must judge you in order to console someone else. God gives mercy to the merciful. Today I would be wise to look at my life and evaluate my standing before God based on my treatment of others. God is a God of mercy. He gives mercy to us when we humble ourselves and when we display that mercy to others.

THOSE WHO GIVE MERCY GET MERCY — DAY 212

Nahum 3:7 And it shall come to pass, that all they that look upon thee shall flee from thee, and say, Nineveh is laid waste: who will bemoan her? whence shall I seek comforters for thee?

What kind of help could you expect tomorrow if you lost all your leverage today? If you had no money and no authority, what kind of help could you expect to receive from people who wanted to give it? You might think on that for just a moment, because some of the most powerful people and nations in this world have built themselves at the expense of others. History is full of the rise and fall of empires, almost all of which came to their zenith by raw power and the leveraging of that power against others.

Assyria, represented by the great city Nineveh, was no exception. She came onto the world stage by sheer force of cruelty, exertion of power, and ambition. Nahum pronounced woe on this city built on the misery and at the expense of others. Thus, when the time came that they were weak and no longer had the ability to force people to aid them, they had no help. The reason should be self-evident. Those who withhold or lightly esteem mercy can expect none in return.

Here was mighty Nineveh who lightly esteemed mercy on both sides. On one side, God had extended mercy to them one hundred years earlier when Jonah had brought a message of judgment. Nineveh had turned to God but then forgot God and very quickly regressed back to their old ways of cruelty and oppression. They lightly esteemed God's mercy, and thus they lightly esteemed the mercy they could have shown to others who were not as powerful as they were.

What do you suppose the reaction of the world was when Assyria was at the mercy of someone else? Nahum 3:7 says, *"And it shall come to pass that all they that look upon thee shall flee from thee, and say, Nineveh is laid waste: who will bemoan her? whence shall I seek comforters for thee?"* There were no comforters for Nineveh, only those who rejoiced in her destruction. The last verse of this book says, *"All that hear the bruit of thee shall clap the hands over thee: for upon whom hath not thy wickedness passed continually?"* They were reaping what they had sown.

Now you're an individual, not a nation, but you and I should both remember that the attributes that characterize our rise to power may also accompany our fall from power. **Those who are merciful receive mercy.** Those who are humble give it, and those who are wise do not take mercy for granted. They thank God for it and then pass it along.

◇◇◇

DAY 213
FAITH: WHAT WE SEE OR WHAT GOD KNOWS?

Habakkuk 2:4 Behold, his soul which is lifted up is not upright in him: but the just shall live by his faith.

Most of us like a good mystery. We're entertained and intrigued by a good mystery, whether it is in a novel or in some news story we're following. But I don't think anyone alive is entertained by the mysteries of his or her own life. No, we are frustrated by such mysteries. All of us can see disparities in life. We see what is and what we believe should be, and we can't help but question. It's a mystery to us how such things could be.

Habakkuk was a man keenly aware of such mysteries. He was surrounded by wicked people who literally seemed to be getting away with murder. He asked God, "How long am I going to watch this wickedness go on with no punishment?" God's answer to Habakkuk seemed to make things worse. Basically, God said, "I am working. I will work, and I'm going to judge the wicked people among your own with the even more godless Babylonians who are going to overtake them."

Now Habakkuk asked new questions, "Well, God, how can you judge wicked people among your own with even more wicked people that are pagan?" This illustrates a couple of the fundamental questions in life: How do we process the mysteries of life when we don't understand? How do we have peace when there are such disparities between what is and what should be?

Well, it's interesting to note Habakkuk's own concession about Who God is, because therein lies our clue to solve mysteries of life. In Habakkuk 1:12 he acknowledges that God is the Holy One, everlasting, self-existent, and mighty. Shifting his focus from what he understood to Who God is, Habakkuk came to a place where he could simply wait on God's answer.

The real answer to all this is found in Habakkuk 2:4, *"Behold, his soul which is lifted up is not upright in him [a reference to the Babylonians' arrogant armies]: but the just shall life by his faith."* The way to negotiate the mysteries of life is to put our faith in what God knows and not what we see. Sight just isn't as reliable as we would like it to be, but God's understanding is infinite.

The key to life is not *understanding* things; it is knowing God. It is putting faith in the character of a God Who is everlasting, all-powerful, and good. It is putting our faith in what God knows and not merely what we see.

DAY 214

GREAT EXPECTATIONS

Zephaniah 1:12 And it shall come to pass at that time, that I will search Jerusalem with candles, and punish the men that are settled on their lees: that say in their heart, The LORD will not do good, neither will he do evil.

One night after a church service, I rode on patrol with a police officer (who was a friend of mine) from the church. While working the night shift, we responded to a domestic call from a father whose son had stormed out of the house in great anger. When we arrived, the teenage son cursed at my friend (the police officer) and angrily walked off. He had nowhere to go, but he just stormed off aimlessly down the city street at midnight.

I vividly remember the father turning to my friend and saying, "He's just given up on life." At seventeen years of age? It sounds unbelievable, but for this young man, life seemed hopeless at only seventeen! If you are taught in school that there is no God, that we are just some cosmic accident, and that man is his own god, then life can seem hopeless and without purpose, even to a seventeen-year-old. Many have low expectations or no expectations.

Zephaniah was a prophet who spoke to people with very low expectations of God. They did not expect much from God, so they did not turn to Him with their whole heart. The small book of Zephaniah at the end of the Old Testament reveals God to be actively searching and judging, for **God is a God of action.** However, the men in Jerusalem were sitting back, not expecting God to do anything.

The truth is that **God does not reveal His power to complacent people.** What do complacent people expect? Nothing. What do they want? Nothing. Complacent people are satisfied. Those who know God should have expectations that are worthy of Him. Any dealing of God in your life is better than no dealing at all!

What are your expectations of God? Do your expectations simply mirror your own personal history and abilities, or do they mirror the scope, magnificence, power, and holy nature of a God that refuses to be limited? How great are your expectations of God?

◇◇◇

WE MAKE TIME FOR OUR PRIORITIES

DAY 215

Haggai 1:2 Thus speaketh the LORD of hosts, saying, This people say, The time is not come, the time that the LORD's house should be built.

Chances are good that you are skimming this. We have become masters of skim reading and skim living. We have very little time, and by every observable indication we seem to be very mindful of our time. Yet, we don't seem to have time for the most important things. Our bodies wear out; our spirits break down; our families tear apart. Oftentimes, these things happen because we're spending time on things that may be good, but not important. People make time for their priorities.

Haggai was God's choice to God's people after the Babylonian exile. By God's command, they had begun rebuilding the Temple; but because of man's opposition, they stopped. Yet, while the opposition stopped them from building God's house, it did not keep them from building beautiful homes for themselves. It's interesting to note that there was no opposition against building their own houses. There was only opposition against building God's house.

So what did God say to these people? Haggai says in Haggai 1:2, *"Thus speaketh the LORD of hosts, saying, This people say, The time is not come, the time that the LORD's house should be built."* They weren't against building God's house: they just didn't think this was the time. It seems as if there's no perfect time for the most important things. In verse 4 God says, *"Is it time for you, O ye, to dwell in your cieled houses and this house* lie waste?... *Consider your ways.*" God is reminding us just as certainly as He was reminding His people of old that you spend time on whatever you value.

The question of value is a "who," "what," and "when" question.

For Whom would they invest their time, effort, and money? The Lord or self?

On What would they spend their time, effort, and money? The Lord's house or their own?

When would they build God's house? Ironically, it wasn't time to build God's house, but they somehow found time for their own houses.

To whom are you giving your time and for what purpose? What are you investing in? First of all, have you trusted Christ? Have you made time for your soul? What about God's Word? Are you spending time with your family? Time is one of the most precious commodities that God gives, and He can give you the guidance to use it wisely if you're willing to accept it. **Let's value the most important things and spend our time accordingly.**

◇◇◇

DAY 216
IF GOD DID, THEN GOD CAN

Haggai 2:3 Who is left among you that saw this house in her first glory? And how do ye see it now? Is it not in your eyes in comparison of it as nothing?

I have to confess to being very sentimental. Now this is my temperament, I suppose, but I also have some of the best of reasons. Chief among them is that God has been good to me. I've had a good life with so many good people and good things, and I fondly look back on these. Of course, my sentimentality is also the reason I have a junk drawer full of, well, junk. My thinking is that maybe these things will become valuable someday, and they certainly remind me of golden memories in my past. But if I'm going to make room in my drawer for things of value, I've got to get rid of the junk.

At any rate, maybe you long for the good old days as well. You're sentimental like me, full of nostalgia and wistfulness for what has been. Now, nostalgia has its place, but nostalgia should never be the enemy of vision. Unfortunately, this can happen. It did in Haggai's day.

The temple that Haggai and others remembered was Solomon's magnificent Temple, one of the wonders of the world. It had been massive in size, gorgeous in sight, and singular in scope, but it had been destroyed. So God's people, by God's command began rebuilding the Temple. Those who had any recollection of the former Temple looked on the new Temple with a little bit of disdain and discouragement. The new Temple was smaller and plainer. Their vision of the future didn't quite match up with their inflated view of the past.

The question was one of perception. How did they see the Temple now? Their standard was a temple that no longer existed and had only grown larger in their memory. Now, sometimes comparisons can be helpful, but oftentimes they can only drag us back and keep us from moving forward.

In Haggai 2:4 God says, *"Yet now be strong… all ye people of the land…and work: for I am with you."* I love that. If you remember nothing else, remember, *"Work: for I am with you."* It's God's work. We're His servants, and He wants to do what needs to be done through us.

Some people try to run from their past, and some people are stuck in the past voluntarily. Allow the past to encourage you that God can, but don't let it convince you that God won't. **If God did, then God can.** Remember that God's work is God's, not yours. The Temple was God's house. But what was most important was not the house; it was the God that inhabited it. The temples no longer remain, but God does. Thank God for His working in the past! Be encouraged by it, and forge on into the future remembering that God is with you when you're with Him.

DAY 217

YOU HAVE THE POWER OF CHOICE

Zechariah 1:3 Therefore say thou unto them, Thus saith the LORD of hosts; Turn ye unto me, saith the LORD of hosts, and I will turn unto you, saith the LORD of hosts.

Just recently, I was meeting people before a Sunday service, when I met a man that had the exact same name as an old family friend that I had not seen for some time. When he told me his name, I stopped for a moment, a bit startled, and asked him to repeat the name. Come to find out, he was the son of our family friend. The more I thought about it, the more it made sense. He looked so much like his dad! We often have the same name, personality, coloring, or facial features as our parents or some ancestor.

For many years in the past, people pretty much did whatever their parents did. Boys, in particular, were expected to adopt the same profession as their fathers. Though we are often similar to our ancestors in appearance or occupation, it's important for every one of us to remember that we have a choice. You see, your ancestry is fixed, but your future is not. **You have the power to choose.**

That choice is clearly articulated by Zechariah in Zechariah 1:2-3, where he says to the people of Israel, "*The LORD hath been sore displeased with your fathers. Therefore say unto them, Thus saith the LORD of hosts; Turn ye unto me, saith the LORD of hosts, and I will turn unto you, saith the LORD of hosts.*" God was talking to His people about the precedent that had gone before them, and He indicated that they had a choice. If they chose to turn toward God, God would match them step for step.

Regardless of what your history or heritage may be, you alone will answer to God for your actions. In verse 4 God tells them, "*Be ye not as your fathers, unto whom the former prophets have cried, saying, Thus saith the LORD of hosts; Turn ye now from your evil ways, and from your evil doings: but they did not hear, nor hearken unto me, saith the LORD.*" The prophets had told their parents to turn from their evil ways, but their parents did not hearken unto God. There is no excuse for following a poor precedent when we have centuries of man's history and God's eternal Word to inform us.

I don't know what path you've been led to or what path you've seen others take, but none of us should mindlessly plow ahead. None of us can change our ancestry or heritage, but each of us is responsible for the decisions we make for our future. You are the link between your family and destruction or the blessings that God intends. The way you choose and the things you do today are pretty important! Rest assured that regardless of your family history, God will be with you and will make a difference through your life, if you will choose His way.

DON'T REPLACE GOD WITH A TRADITION

DAY 218

Zechariah 7:6 And when ye did eat, and when ye did drink, did not ye eat for yourselves, and drink for yourselves?

Suppose you run into a friend who tells you that he has just celebrated his 25th wedding anniversary by gathering some friends to have a big dinner, play putt-putt golf, and stay out late. You are shocked, and you ask him what his spouse thought about it. Your friend replies, "My wife didn't care. She celebrated with a quiet evening at home alone." Now that would be pathetic and laughable. You're not celebrating an anniversary unless you are remembering the one to whom you're married.

God had instituted a number of fasts, feasts, and traditions for His people Israel, recorded in the Old Testament. God's people had also instituted some traditions of their own. None of these traditions were bad; they were good. But the point of all these traditions was to point God's people toward Him; the point was not for the traditions to become a replacement for Him.

Zechariah 7 speaks of a tradition that commemorated the burning of the first Temple by Nebuchadnezzar in 586 BC. When

God's people began construction of the new Temple, they began to wonder if they should still be commemorating the burning of the first Temple.

They brought this question before God Who responded with two probing questions. The first question is found in Zechariah 7:5, *"When ye fasted and mourned in the fifth and seventh month, even those seventy years, did ye at all fast unto me, even to me?"* Then verse 6 says, *"And when ye did eat, and when ye did drink, did not ye eat for yourselves, and drink for yourselves?"* Notice the contrast between a commemoration that is unto God and one that is for ourselves.

God is not saying that tradition is wrong, and God wasn't necessarily even condemning the memorials themselves. When we practice traditions in our lives, we need to ask two questions: "What is the purpose?" and "Are we fulfilling that purpose?" Tradition is good, and it is a part of the rhythm of life, but **God is unimpressed by any tradition that takes the place of loving Him and others.**

◇◇◇

DAY 219
IN DUE SEASON

Zechariah 10:1 Ask ye of the LORD rain in the time of the latter rain, so the LORD shall make bright clouds, and give them showers of rain, to every one grass in the field.

Do you remember one thing you prayed for last fall? Whatever you prayed for, you either kept asking, stopped asking, or asked someone else. You may have kept asking because some needs are continual like rain, wisdom, finances, and health. You may have stopped asking because you no longer had a need, you didn't remember your need, or you no longer cared. You may have asked someone else because you still cared but no longer believed that God could or would answer.

Everything you need is supplied from heaven and in due season. That is why the Bible encourages us that *"in due season we shall reap, if we faint not."* (Galatians 6:9) The children of Israel had languished in building the Temple after twelve long years. In response, God said, *"Ask ye of the LORD rain in the time of the latter rain; so the LORD shall make bright clouds, and give them showers of rain, to every one grass in the field."*

I remember asking God for rain here on the Ranch because we desperately needed it. I remember God answering that request with several inches of soaking rain! Just because we asked God for rain in October does not mean that we cannot or should not ask for rain in the spring or summer. You should not expect to pray one time and be set for life. Remember, every blessing comes in due season.

You may have needs that you have prayed about before, but if they are needs now, *"ask ye of the LORD."* **Don't try to live today on God's provision from last fall!** Perhaps you may be tempted to stop asking or to ask someone else, but *"ask ye of the LORD."* Remember that every blessing comes *from God* in *due season*. Just like the former rain that is past and the latter rain that we need right now, *"ask ye of the LORD"*—and keep asking the Lord—for what you need!

PERFECT PERSPECTIVE

DAY 220

Zechariah 14:9 And the LORD shall be king over all the earth: in that day shall there be one LORD, and his name one.

Have you ever been to New York City or some other huge city? In a city like New York, the difference in perspective from the subway to street level to an observation deck in a tall skyscraper can be pretty amazing. You can be in the same physical city, but your perspective makes all the difference.

There is no difference between the ancient past and the distant future to God. God lives in the eternal present. He is always present in time and space, and His "sight" is 20/20. He sees the past and the future in perfect clarity. If we are not careful, we can live on the latest breaking news and become so wrapped up in today that we lose perspective of the eternal.

The truth is, there is coming a day when the Lord will establish His kingdom. He reigned in Zechariah's day; He reigns in our day; and He will reign forever and ever. In light of the fact that one day, *"the LORD shall be king over all the earth,"* we ought to **live in today** and **look to tomorrow**. Both are necessary for a proper perspective. Don't live only in today with no thought for tomorrow; on the other hand, don't squander today because you are only looking at tomorrow.

Today is the only day you have! You can't change yesterday, and you are not guaranteed tomorrow. What you can do is what you ought to do right now. On the flip side, don't sacrifice the important for the immediate.

Perspective does matter, and no one has a better perspective than God Almighty. The problems of life that are way over your head are way under His feet. God reigns, and one day He will be King over all the earth. If you will live in today and look to tomorrow, you can have the right perspective on what is and what is to come.

"The kingdoms of this world are become the kingdoms of our Lord, and of his Christ; and he shall reign for ever and ever." (Revelation 11:15)

DAY 221

SELF-ABSORBED AND SHORT-SIGHTED

Malachi 1:2, 6 ...Wherein hast thou loved us?... And ye say, Wherein have we despised thy name?

This is mean to say, but let's face it, absent-minded people are fun to watch. Maybe you've known a friend who was so absent-minded that he could walk right off the end of a pier and not even know what hit him. I can be absent-minded myself. Some time ago I was driving down to Florida, and I was so wrapped up in what I was thinking about that I missed my exit by about 50 miles. It's a hard life for absent-minded people! But at the end of the day, an absent-minded person does have a mind. It's just that the mind is not present. It is unfocused.

When you read the book of Malachi you can't help but get the impression that the people to whom God is speaking, His own precious people, were absent-minded. They didn't know what was going on because they were self-absorbed. They were so focused on themselves and their own desires, problems, and disappointments, that they totally missed the larger picture and the great blessings that God had given them.

First, these **self-absorbed people didn't see how God had blessed them.** In Malachi 1:2, God professes His love for Israel, and they respond by saying, *"Wherein hast thou loved us?"* They seemed to doubt the word of God that He loved them. They were so absorbed in their own world that they didn't see how God had blessed them. This blindness that defined a people can also define an individual.

So many times I'm so absorbed with my problems that I totally miss the ways in which God has blessed me, taken care of me, and provided for me. At such times I'm walking through life blind and short-sighted.

Secondly, **self-absorbed people do not see how they're ignoring God.** In verse 6 they asked God, *"Wherein have we despised thy name?"* God had addressed the priests in particular and pointed out how they had treated God lightly. They had lightly esteemed Him, given their second best, and certainly not given Him their hearts.

Do I fail to see how I may be lightly esteeming God? Do I fail to see how God has greatly blessed me? Am I more absorbed with how impressive I look or how comfortable I feel than I am about what God thinks of me? If I'm going to honor God, I need to recognize Who He is, be mindful of how He has blessed me, and be conscious of how I esteem Him.

DAY 222

REVEALING GOD

Malachi 1:13 Ye said also, Behold, what a weariness is it! and ye have snuffed at it, saith the LORD of hosts; and ye brought that which was torn, and the lame, and the sick; thus ye brought an offering: should I accept this of your hand? saith the LORD.

Do you remember playing hide-and-seek? You probably played it as a child, but have you ever played the game with kids? Sometimes they hide by simply putting their hands over their eyes. They figure if they can't see you, you can't see them!

That perspective is childish and funny for a game like hide-and-seek, but God's people had the same attitude towards the Lord in Malachi 1. Since they could not see God, they thought somehow He could not see them. Is God unable to see or hear? Is He clueless when it comes to life and the world around us? To even ask the question is silly, but that was how they treated God!

How highly you esteem God boils down to your attitude and your actions. **Your attitude and your actions reveal to others a God that no one has seen.** No one you know has actually seen God; but everyone you know has made some judgment about God by your attitude and actions. What does the caliber of your work say about your God to those around you?

Reading through the first chapter of Malachi, you will catch the wrong attitude of the children of Israel. They were mostly **characterized by questioning.** Now asking questions is not wrong, but there is no more effective tool in a rebel's toolbox than the tool of questioning. You can never satisfy a rebel's questions!

Not only were their attitudes rebellious, Israel's sacrifices were inferior. They were bringing sacrifices that were torn, sickly, and lame. They **brought less than God deserved and were insincere in their giving.** God's reply was, *"Should I accept this of your* *hand?"* He is the Lord of hosts and deserves the very best, but their actions did not reveal that truth.

The question to ask yourself this morning is this: **Would the perfect Master be pleased with the caliber of my work and my attitude today?** What do your attitudes and actions reveal about your God? Is He getting your very best? Would a lost world have a high estimation of God by watching you?

KNOWLEDGE IS WORTHLESS WITHOUT OBEDIENCE

DAY 223

Matthew 2:4 And when he had gathered all the chief priests and scribes of the people together, he demanded of them where Christ should be born.

Every time I start reading in the New Testament I am reminded of how much of the Bible is Old Testament. The Old Testament is huge, and it is densely populated with God's truth. However, all of the Old Testament culminates in Matthew 1 in the person of Jesus Christ. The very first words of Matthew are, *"The book of the generation of Jesus Christ."* This last genealogy of the Bible illustrates that all the genealogies of the Bible culminate in one person, the Lord Jesus.

If you scan Matthew 1-3 you find a repetition of the idea that God is fulfilling prophecy through His Son, the Lord Jesus. For instance Matthew 2:15 says, *"That it might be fulfilled…."* Verse 17 says, *"Then was fulfilled that which was spoken by Jeremy the prophet…."* Again in Matthew 2:23, *"That it might be fulfilled…."* Matthew's point is that the birth of Jesus Christ was according to God's plan.

Now of all the prophecies mentioned in the first part of Matthew the one that strikes me the most is the one recited by the chief priest and scribes when King Herod demanded of them where Christ should be born. Their answer to Herod was, *"For thus it is written by the prophet."* These Bible scholars knew prophecy, understood in some measure God's promises, but totally missed God's Son. Why? **Because an understanding of prophecy is worthless without a willingness to obey God in the present.**

Can I tell you something? You know everything you need to know in order to do what God wants you to do today. You don't have to know the future to know what is right today because so much of what God wants today is just obedience to the basic principles of His Word. Can God show us things that are more specific and detailed? Of course! But it begins, not with a fascination of the future, but with a willingness to obey God right now. **The key is not to know the future, but to know the God Who knows the future.** Nothing is more important than obeying God right now so that you can be where you should be tomorrow.

◇◇◇

DAY 224 — THE POWER OF GOD'S TRUTH

Matthew 4:1 Then was Jesus led up of the spirit into the wilderness to be tempted of the devil.

Do you know how to make a dead man talk? You probably do, and most of us have done such a thing. Most of us have quoted people who said something brilliant but are now long gone. It is interesting to hear two people on opposite sides of an argument both quoting the same person that they both revere. Of course, the person they are quoting is not alive to clarify exactly what he meant. There is no doubt, though, that words have power; and words have even more power when they are true. No words are more powerful than the ones that God has given us in Scripture.

Matthew 4 emphasizes this in an unusual way because here we find the temptation of the Lord Jesus Christ by the devil himself. The Bible tells us that the tempter came to Jesus and said, *"If thou be the Son of God, command that these stones be made bread."* Now much could be said about this temptation and the devil's ploy. What I want to emphasize is how Christ answered. He said,

"It is written…." Again in verse 7 Jesus answered a temptation by saying, *"It is written…."* Again in verse 10, Jesus answered the devil by saying, *"It is written…."* It is remarkable to me that even Christ used Scripture in response to temptation.

Now, if the Lord Jesus put that kind of high regard on the Word of God, surely you and I can't afford to think that we're empowered and enabled to stand up to the devil without submitting to, understanding, and being saturated by the Word of God.

What is even more fascinating is that the devil quoted Scripture. In verse 6 he quotes, *"As it is written…."* Now when Christ uses Scripture, that is power. But when the devil uses Scripture, it is a fraud. The devil is never submissive to the Word of God; and he twists it for his own purposes.

The Word of God is powerful for those who live in submission to it. God's Word is not a magic charm that has some power other than its meaning. What makes God's Word powerful is that it is true; and it has power in my life when I submit to it. That means I need to do more than just read it or quote it. I need to apply integrity by connecting God's Word to my life.

God's Word is powerful, but it is powerful to people who live it honestly and read it submissively. Even the Lord Jesus used the Scripture. He used it powerfully because He used it accurately. Even the devil can quote Scripture, but **the power of Scripture is unleashed in my life when I read it honestly and submit to it in my life.**

◇◇◇

BEFORE MEN

DAY 225

Matthew 5:16 Let your light so shine before men, that they may see your good works, and glorify your Father which is in heaven.

In the Sermon on the Mount, the Lord Jesus said, *"Let your light so shine before men, that they may see your good works, and glorify your Father which is in heaven."* We are to let our light shine **before men**, so that others can see, so that they will glorify God. Then, in the same sermon, the Lord Jesus said, *"Take heed that ye do not your alms before men, to be seen of them...."* We are not to do our alms **before men** so that they see our good works. Confused yet?

Which is correct? We are talking about two points in the same sermon, at the same time, made by the same Speaker! How do you reconcile letting your light shine so that men can see your good works but not doing your alms (a good work) before men? Well, the answer lies in one word: **motive**. If your motive is glorifying yourself, like the hypocrites did in the synagogue, you are missing the point!

When you let your light shine, others should be impressed with God's glory, not with how "bright" you are. The context of Matthew 5:16 is persecution (see verse 11); the context of Matthew 6:1 is praise of men. The truth is, if you are out to gain the praise of men, your light will probably be the "dimmest" when persecution comes. Some of the loudest voices at church on Sunday are the quietest voices in the workplace Monday through Friday.

Your life can either glorify you, the world, or God. It's not possible to glorify all three at the same time. If you are getting credit for how right and wonderful you are, God is not receiving His due glory. So today, let your light shine before men, in order that God will receive the credit. Who is getting the credit in your life?

◇◇◇

DAY 226: PLAN LONG BUT WORRY SHORT

Matthew 6:34 Take therefore no thought for the morrow: for the morrow shall take thought for the things of itself. Sufficient unto the day is the evil thereof.

If you were to take all of your possessions and stack them in one place, how much space would it require? Well, for some of us it might require a postage stamp. For others, it might require a pole barn or maybe a hangar. What about Richard Branson, the billionaire entrepreneur? He owns a number of ventures and businesses. I wonder how much space it would require to hold all that he owns. There's a sense in which the space it would take to hold all of your stuff and the space it would take to hold all of Richard Branson's stuff is the same space. It's the little 24-hour time period called "today."

Whatever problems or possessions you own, you're going to hold them today. You don't own tomorrow, and you've lost yesterday. All you have is today. Yet, so many times we're planning so far ahead that we worry and cram all of tomorrow's worries into this little space called today. I'm not against planning, and I don't think God is either. But one of the dangers of planning is to substitute planning for trusting God. One of the side effects of planning can be to drag tomorrow's problems into today. You can regret the past, but you can only worry about the future: we often do.

In Matthew 6 Jesus is talking about worry, priorities, and putting God and His kingdom first. In Matthew 6:34 He says, *"Take therefore no thought [anxious worry] for the morrow: for the morrow shall take thought for the things of itself. Sufficient unto the day is the evil thereof."* God's grace is sufficient to this day. The problems that God allows for this day will not overflow this day. We may feel like they do, but they do not because God's grace is sufficient.

In Matthew 6:11 Jesus gives us a model prayer, and He says, *"Give us this day our daily bread."* Most of us have never prayed for daily bread because we have bread for today, tomorrow, and the next day. "How will I have bread for the next forty-five years?" we worry.

John Rice used to say, "I live a hand to mouth existence: God's hand to my mouth." Now, I'm not advocating carelessness in a sense of being lazy and not planning, and neither is God. But the Lord Jesus is telling you that your problems will not overflow this day and that God's grace is sufficient for this day and in this day.

DAY 227

THE GOLDEN RULE

Matthew 7:12 Therefore all things whatsoever ye would that men should do to you, do ye even so to them: for this is the law and the prophets.

Do you realize how often we think about "them"? You and I spend a big portion of our days thinking about "them"—our students, our customers, our friends, our family. If we're not careful, we spend all that time thinking about our duty and sometimes resenting it. I'd like to give you permission to make this a "you" day. That's right, we're officially declaring that today is You Day! Now there's one little caveat to that: you are free to think about what you would have other people do for you, and then go do those things for others.

In Matthew 7:12 the Bible records Jesus saying, *"Therefore all things whatsoever ye would that men should do to you, do ye even so to them: for this is the law and the prophets."* We often call this the golden rule, and it is found in other variations throughout literature. It has been quoted for centuries in various religions. This tends to highlight the essential nature of what God has mandated and what Jesus has said.

Can you imagine how different your day would be if you simply treated everyone the way you would want to be treated? Too often we wait until some rule is made, and we're forced to do something we should be doing anyway—treating people with what might be called common decency. The irony is that common decency is sometimes anything but common. We need to remember that we will reap what we sow. If everyone in my world is sarcastic, or stingy, or guarded with me, it may well be a mirror of what I am being to them. Sometimes a man is the last person to see himself as he really is.

The wonderful truth is that **if I will put others first by treating them the way I would want them to treat me, I will reap blessings in my life and be a blessing to others.**

YOU DON'T HAVE TO HAVE ALL THE ANSWERS

DAY 228

Matthew 11:3 And said unto him, Art thou he that should come, or do we look for another?

Do you remember your hardest class in school? Maybe it was math, English, or history. Most of us remember what it was like to be in a class that was a challenge to us. On occasion I've been in a class where I felt like I was the only one who didn't understand what was going on. I had so many questions, but I didn't ask any of them because I didn't want everyone to know just how dumb I was! Well, if you remember that feeling, you may also remember anticipating the day when you'd be a smart adult and have all the answers. But if you are an adult, you know that life still has the capacity to raise questions that you cannot answer.

John the Baptist knew exactly what that was like. John had some raging questions that were brought on because everything in John's life was going wrong. John was in prison and would eventually be beheaded. All that John had expected, sought, and preached didn't seem to jive with what was actually happening in his life. So John did a very bold and humble thing. He sent people to ask Jesus.

Verse 2 says, *"Now when John had heard in the prison the works of Christ, he sent two of his disciples, and said unto him, Art thou he that should come, or do we look for another?"* John had this question because he was having a hard time. I love the way Jesus responds. Jesus told them, *"Go and shew…."* He didn't send them back with an argument. He sent them with something they could see. Now what were they to show? Were they to show some theological truth, or were they to show some practical experience? The answer is yes. Both are embodied in the same person, Jesus Christ. He told them, *"Go and shew John again those things which ye do hear and see."*

Basically, Jesus was appealing to the knowledge of John. John knew what the prophets of old had prophesied that the Messiah would do. He was sending them back to tell John what they'd been seeing. They were going as witnesses to tell what they knew. John would know that Jesus was exactly Who the Holy Scriptures said the Messiah was to be.

That same assurance is available to you today. When you are facing doubts, do what John did and turn to face the Lord. The Lord Jesus is not put off by questions from a submissive, honest heart. You don't have to have all the answers. You just have to have the wisdom to know where to look. You are free to acknowledge that you don't have all the answers, so be willing to ask the One Who does.

◇◇◇

DAY 229 — WHO OR WHAT?

Matthew 13:58 And he did not many mighty works there because of their unbelief.

It is quite interesting that the Lord Jesus would literally limit Himself according to the belief of the people with whom He was working. Would Christ have done more mighty works than He did if they had believed? Certainly, He would have. What was it that they did not believe? They did not believe *Who* He was.

In verse 54, the people said of Jesus, *"Whence hath this man this wisdom, and these mighty works?"* They believed that the Lord Jesus could do mighty works—they had seen Him do miracles. But they questioned where the mighty works came from. **Their problem was not the** what; **their problem was the** Who. They saw the works, but they were amazed that the works would come from *"the carpenter's son."* (verse 55)

Sometimes we can see "mighty works" and attribute them to our abilities or still doubt the One who did them. ***Who* you believe is more important than *what* you believe.** In other words, believing in *Christ* is more important than believing what Christ can *do*.

Who you believe determines what you believe. There is a difference between having confidence that something will happen and having confidence that God is good and will take care of you. What happens if I really have confidence in something happening, but it never comes? In that case, I am still unshaken if I believe the Person behind it all.

There is peace in the Christian life when you trust God (the Who) instead of the thing (the what). No matter what comes or doesn't come, all will be well because God is good. These folks in Matthew 13 literally overlooked the Lord Jesus because He was too close. They took him for granted because He (God in flesh) was so close.

Don't fix your attention on the "things" you need or expect to come. Those "things" are only as good as the person delivering them. The place to start is to believe God; and no matter what may come or not come, you can know it will be good because the "Who" behind it is good!

◇◇◇

GOD'S MULTIPLICATION

DAY 230

Matthew 14:16 But Jesus said unto them, They need not depart; give ye them to eat.

Chances are good that your day is consumed by feeding hungry mouths. If you're a parent, a mother in particular, so much of the time is spent getting the food, preparing the food, eating the food, and then repeating the process to keep kids and a husband fed. If you're in an office, so much of your day is defined by deadlines, projects, and people that want what you have now. If you're a teacher, you are filling little minds with what they need in order to negotiate life. If you work with your hands in some way, there are often times when you feel like you need a hand; you need help. If you're a preacher, you are daily in need of "bread" to give your brothers and sisters. There are hungry mouths to feed.

You know, the Lord Jesus knows exactly what that is about. In Matthew 14, He found Himself surrounded by a great multitude in a desert place in the evening. His disciples said, "Lord, send these people away. They need something to eat." The Lord's response is remarkable. In verse 16 He said, *"They need not depart; give ye them to eat."* His answer almost seems to be casual. It was absurd on the face of it to think that these disciples could feed this multitude. It's also obvious, at least in hindsight, that the Lord Jesus could.

They found that they had five loaves and two fishes. What was that compared to the need! In a world defined by supply and demand, the demand was far greater than the apparent supply. The answer for this is found in verse 18 where Jesus said, *"Bring them [the bread and fish] hither to me."*

What was only five loaves and two fishes in the hands of the disciples was a feast for five thousand in the hands of God the Son. The Bible says in verse 20 that when He had fed the multitude, "they did all eat, and were filled." In the end, they had more left over than they had when they began!

What is the lesson from all of this? The Lord can certainly provide miraculously, but notice that Jesus provided miraculously by multiplying what the disciples gave Him. **Sometimes, the Lord is not as interested in giving you something amazing and new as He is in multiplying what you give Him.**

When you are stretched between a great demand and a meager supply, there are two things you need to do. You need to use what you have, and you need to ask God for more. Now, you may only have the equivalent of five loaves and two fishes when it comes to time, money, or know-how; but submit what little you have to God, and let Him multiply it for the benefit of others and for His own glory.

Then, ask God for more. There is nothing wrong with asking God for more when you have been a good steward of what you already have. God takes what we have, which is what He gave to us to begin with, and multiplies it in ways we can't imagine and in ways that bring glory to Himself.

PAY ATTENTION!

DAY 231

Matthew 15:33 And his disciples say unto him, Whence should we have so much bread in the wilderness, as to fill so great a multitude?

Do you find it hard to pay attention? You probably do. If you're like most modern people, it is harder for you to pay attention now than it was twenty years ago! In fact, it may be harder to pay attention now than it was twenty minutes ago. There are a variety of reasons for this, but I will tell you that many people we consider brilliant probably are not any smarter than you or me. They just pay better attention than we do. They learn from life and from God, and they put these lessons into practice in ways that we don't because we are too oblivious to what's happening around us.

Now if you have a hard time paying attention, you will probably sympathize with the frail memory of Jesus' disciples in the story of Matthew 15. Jesus had just recently fed a hungry multitude, and He and His disciples were again confronted with the same situation. When, for the second time, the disciples realized that there was a huge crowd with very little food, they said unto Him, *"Whence should we have so much bread in the wilderness, as to fill so great a multitude?"* (Matthew 15:33)

Right after this second feeding of the multitude, the disciples were in a boat with Jesus and were concerned that there was not enough food for the thirteen of them. The disciples were worried about having enough food right after He'd fed thousands of people right before their very eyes!

It's pretty clear that the disciples were not paying attention. The disciples failed to learn from the Lord and from life. Because they had not been paying attention the first time the Lord fed a multitude, they doubted that He could provide the second time they faced the same situation. We can learn from their mistake.

The lessons that you learn need to be measured by what God has said, so that they are accurate. If you make a mistake once, that's common; but if you make the same mistake over and again, that is needless pain and trouble. **Pay attention. Learn from the Lord. Learn from life**, and you'll be a blessing.

◇◇◇

DAY 232
THE SOURCE OF WHAT YOU SAY

Matthew 16:17,23 And Jesus answered and said unto him, Blessed art thou, Simon Bar-jona: for flesh and blood hath not revealed it unto thee, but my Father which is in heaven... But he turned, and said unto Peter, Get thee behind me, Satan: thou art an offence unto me: for thou savourest not the things that be of God, but those that be of men.

We all know someone who "talks out of both sides of his mouth" or is "two-faced." What concerns me more is when two different things come out of my own mouth! Isn't it amazing how often we battle between what we *should* say and what we *do* say? Matthew 16 reminds us that there are two different sources for what we say.

Jesus asked His disciples, *"Whom do men say that I the Son of man am?"* Of course, there were a variety of answers. Then He asked, *"But whom say ye that I am?"* Peter actually answered correctly in verse 17! However, did the Lord praise Peter for being the smartest disciple or being brilliant? No! He said, *"My Father which is in heaven* [hath revealed it unto thee]." God gave Peter the answer. Peter finally got one right, yet he got none of the credit!

Only a few verses later, Peter rejected the Lord's mission for coming to earth. While Peter was encouraged when he answered the question of who Jesus was, he was rebuked when he denied the reason Jesus came. Christ spoke to Peter directly in verse 23, but He rebuked the devil for Peter's words. In both cases, Peter did not get full credit because out of the same mouth came two completely different answers! Yes, Peter was still responsible for his words; but **in one case God inspired his words, and in the other case Satan inspired his words.**

What is coming out of your mouth? Is it sweet water or bitter water (James 3:11)? The devil will not inspire you to lay down your life, to pick up your cross, and to follow Christ. I am reminded (more times than I care to admit) that I say things that I don't want to say. Other times, when I say something good, I am tempted to think, "Boy, that was good!" The truth is, **nothing worthwhile has come from my mouth that came naturally.**

You are never just speaking for yourself. If you find yourself speaking for yourself, that means you are speaking for the enemy. He is the one who works through our sinful human nature. **Every day you are speaking for somebody.** Everything you say is coming from a source. In both cases in Matthew 16, Peter said what he said because he thought what he thought. On one occasion, the Lord was influencing his thinking. On the other occasion, the devil was using Peter's own selfish nature—the same nature that would later swear, deny Christ, and follow afar off.

When you speak this morning, you will speak from a source and for somebody. Who is influencing your thinking? Who is the source for what you say?

◇◇◇

IS GOD AMAZED?

DAY 233

Matthew 24:1 And Jesus went out, and departed from the temple: and his disciples came to him for to shew him the buildings of the temple.

We are living in a day where everyone wants to share every experience and thought with everyone else. How we relate to one another, the speed at which we communicate, and the platform on which we make our individual "voices" heard has completely changed over the last few years. We want everyone else to be amazed and interested in what we are doing.

The truth is, God cares about you, but **He is never amazed at your experiences.** Can you imagine what it would take to impress God? In Matthew 24, the disciples tried to impress the Lord Jesus with what amazed them. Like the Temple they pointed out to Jesus, the things that amaze us are short-lived.

We love sharing our amazement with life's small things; **Christ loves to show us His power and stability throughout life.** The Lord Jesus wasn't amazed by what was, and nothing in your day will amaze Him. Events may surprise you or catch you off guard, but the Lord is never surprised.

Not only that, the Lord is never fearful of what will be. The disciples asked about the sign of Christ's coming and the end of the world. In response, the Lord said, *"See that ye be not troubled...."* Why shouldn't we be troubled? Because the Lord already knows and is in control!

Be ready, be faithful, and trust the One who has power and stability over life. There is nothing wrong with sharing events, thoughts, and stories from our lives that amaze us; but always remember that God is never amazed with what amazes us!

◇◇◇

DAY 234

GOOD WORK OR WASTE?

Matthew 26:10 When Jesus understood it, he said unto them, Why trouble ye the woman? for she hath wrought a good work upon me.

Will you waste this day, or will you invest it? Nearly two thousand years ago, a woman named Mary broke open an expensive box of ointment and anointed the Lord Jesus. You would think this act would have been widely accepted and applauded, especially by the Lord's own disciples. On the contrary, the disciples *"had indignation, saying, To what purpose* *is this waste?"* The closest people to Jesus' earthly ministry were the very people who looked down on Mary's investment.

Sometimes when you give for the Lord and to other people, you will not be appreciated by everyone, even those who are serving the Lord. Some of the best people you know may not understand the choices you make for the Lord. They may characterize what you do as a "waste," but you need not be understood by everyone. You can never please everyone!

Two things, and two things alone, matter when it comes to whether you are wasting or investing: **the Lord's appraisal and His approval.** What Judas called a "waste," Jesus called a "good work." The Lord's estimation of Mary's investment was that it was good. And the Lord's estimation of Mary was that she had done what she could do. (Mark 14:8)

What people see and call a waste, the Lord understands and calls good. **People see what you do; God understands why you do it.** Both sides are important, but one is more important. Do you have the Lord's appraisal and approval for what you are doing today? That is all that matters! Be most concerned with what the Lord understands and knows, not what people think and see.

◇◇◇

LEAVING AND FOLLOWING

DAY 235

Mark 1:18 And straightway they forsook their nets, and followed him.

My grandfather put a borrowed saddle on a half-broke mare named "Go West" in 1930. He was only eighteen years old, but both his parents were dead. All he knew was West Texas; all he loved was ranching. On that day in 1930, he said goodbye to the ranch, cattle, horses, and cowboys, and saddled up to go to a little Bible college in Decatur, Texas. The rest, as they say, is history.

Anyone who has ever been used by God to make a difference in this world had something to leave and Someone to follow.

In Mark 1, the Lord Jesus called His disciples to follow Him. His calling was simple: *"Come ye after me, and I will make you to become fishers of men."* Simon and Andrew *"straightway... forsook their nets, and followed him."* Soon after, Jesus called James and John, and what did they do? *"They left their father Zebedee in the ship with the hired servants, and went after him."* Notice that they all left something in order to follow Someone.

Many people want to follow the Lord but hold on to what they have. You can't do that! While it's not virtuous to be destitute and flat broke, it is virtuous to follow the One who made you. The way that happens is for you to be willing to leave whatever would keep you from following Christ. **Leaving is a means to following.**

So how should we follow? The Bible says that Andrew, Simon Peter, James, and John all followed *"straightway."* That means without delay! Now is the time. You don't have to know exactly what the Lord has planned for the next thirty years in order to follow Him straightway. It's not a matter of where you are going; it is a matter of Who you are following. The Lord knows the way; so if you are following Him, you'll make it!

We should follow straightway, and we should follow every day. The Lord Jesus said, *"I will make you to become...."* That implies some time and some learning. Following Christ is a daily decision.

Follow straightway. Follow every day. And follow all the way! These disciples literally changed the world for Christ because they left something and followed Someone. Don't fall short of all that God has for you because you don't follow all the way. Following Christ is the highest calling you will ever have, and to follow Him you must first leave something. Follow without delay, every day, all the way, and leave whatever may hold you back.

FAITH AND WORKS

DAY 236

Mark 2:5,8 When Jesus saw their faith, he said unto the sick of the palsy, Son, thy sins be forgiven thee… And immediately when Jesus perceived in his spirit that they so reasoned within themselves, he said unto them, Why reason ye these things in your hearts?

This is a great story! A man *"sick of the palsy"* had a need, and his four friends brought him to Jesus. The difficulty came when the crowd was too large to carry this sick man to the Lord, but these four friends weren't discouraged. They went to work on the roof and lowered their friend through the hole they had made!

Verse 6 tells us that some of the scribes reasoned in their hearts—they did not speak out loud. They were critical of the Lord Jesus and thought to themselves, *"Why doth this man thus speak blasphemies? who can forgive sins but God only?"* It is true that only God can forgive sins—no pastor, priest, or person can do that—but they obviously did not believe that Jesus Christ was *"God only."* However, the Lord Jesus *"perceived in his spirit that they so reasoned within themselves…."* **How did Jesus know what they were thinking? Because He is God!**

With that in mind, what caught my attention is found in verse 5: *"When Jesus saw their* [the four friends'] *faith, he said unto the sick of the palsy…."* **How did Jesus see their faith?** Certainly, Jesus is God, but He saw their faith **because of their works described in verses 1-4!** They were not discouraged. They tore the roof off, and they lowered their friend down. That is how Jesus saw their faith.

Faith and works are not the same thing, although they are related. If these friends had depended on tearing off a roof for the man to be healed, they would have been relying on works. But if tearing off the roof and lowering their friend down was a result of their faith, that is something altogether different! Jesus perceived the scribes' hearts because He is God; He saw the friends' faith because of their works. We ought remember today that **our actions need to be driven by our faith.**

WIL RICE IV

DAY 237

THE HEALTH TO SERVE

Mark 1:30 But Simon's wife's mother lay sick of a fever, and anon they tell him of her.

I suspect that we read of the person who had the hardest job in the entire Bible in Mark 1. She didn't command an army; she didn't rule a country; she didn't have some hazardous job like fishing on the tempestuous Sea of Galilee. No, she had a son-in-law who did, and he was just as tempestuous as the sea on which he fished. Of course, we're talking of Simon Peter's mother-in-law. I don't seriously believe she had the hardest job in the Bible, but can you imagine being an in-law to Simon Peter? It certainly wasn't boring or dull.

In Mark 1 we read of Jesus healing Simon Peter's mother-in-law. Jesus came into the house of Simon Peter where Simon's wife's mother lay sick of a fever. Luke, a physician, tells us it was a "great fever" which indicated that this was a serious illness.

The Bible says that immediately after they told Jesus about her condition, He "came and took her by the hand, and lifted her up; and immediately the fever left her, and she ministered unto them." It is interesting to note that Jesus healed her immediately.

It is also interesting that Jesus healed her completely. We know this because she served or ministered to them right away. Now what did she serve? I rather suspect that she was serving a meal. You wouldn't want someone preparing your meal who was only partly over some dread disease, would you? Well, no. So Jesus healed her completely.

Now we have to ask ourselves, why did Jesus heal this woman immediately and completely? I think there's a reason found in the next phrase after her healing. The Bible says that "she ministered unto them." Jesus healed her, and she served others.

Serving others is a sign of health. Serving self is a sign of lack of health. Notice what happened here. He healed: she served. The bottom line is that the moment Jesus healed this woman, she was serving others. Ask God to help you and to give you health to the point that you can serve others. The sign of a healthy heart, a healthy spirit, and a healthy life is serving others, and that's one of the reasons that God gives you breath today.

GOD DOESN'T PANIC

DAY 238

Mark 4:39 And he arose, and rebuked the wind, and said unto the sea, Peace, be still. And the wind ceased, and there was a great calm.

I once read a sign in an office that read, "Poor planning on your part does not equal an emergency on mine." Sometimes, when we have an emergency, the way that people express care or concern for us is by being panicked themselves. They care enough about our situation that they spin their wheels, seem to be in a hurry, and go through a flurry of activity whether they can actually help us or not.

Sometimes, we wonder why God is not in such a hurry when our "emergencies" come. If you read stories of the Lord Jesus, you struggle to find any stories of Him being rushed or in a hurry. In fact, many stories seem to indicate that He actually slowed things down at the actual moment when things were most desperate. The Lord is never frantic or hurried, and He's never worried. He's never at a loss for what to do. He's never out of time. Now don't let that discourage you, and don't ever make that a reason to believe that the Lord doesn't care about you. **The lack of panic on the Lord's part does not equal a lack of concern.**

When Jesus went with His disciples onto the Sea of Galilee, they found themselves in a horrible storm. The Bible tells us that Jesus was asleep, and He wasn't panicked when He woke up. In fact, He was so relaxed in the midst of this storm, that the disciples asked Him, *"Carest thou not that we perish?"* Now why did they make this assumption that He didn't care? They assumed this because He was asleep and not panicked by their trouble.

We learn two things from this story. The first we find in verse 36, *"And when they had sent away the multitude, they took him even as he was in the ship."* So Jesus was with them when they were in this predicament. I can tell you that no matter what you can see or how stormy things may be, Jesus is with you.

Secondly, we learn that Jesus gave them His word. In verse 35 Jesus said, *"Let us pass over unto the other side."* They had His implicit promise. He's with you, and He's able to get you to the other side. I don't know what the other side may be for you today, but remember that Christ is with you. When you read a promise in God's Word, you can trust it. You can rest assured that He cares about you, so don't panic. Trust.

◇◇◇

DAY 239

WHO IS THE GREATEST?

Mark 9:34 But they held their peace: for by the way they had disputed among themselves, who should be the greatest.

Have you ever had a discussion about who is the greatest? Whether you were discussing who was the greatest dad, the greatest player in a particular sport, or the greatest automobile manufacturer, you have probably been involved in a "greatest" discussion. Usually these conversations are hotly contested and become contentious in no time!

For instance, which computer is the greatest, Mac or PC? Who makes the greatest truck, Ford or Chevy? Who is the greatest team in baseball? Football? Who is the greatest player to ever play the sport you love? I could divide my audience in half right here, right now!

Lest you think people only divide over silly things like cars and sports, remember that Christians today can be just as divisive arguing about who is the greatest preacher, what is the greatest Bible college, or what is the greatest church in America. The truth is, anytime the question of "the greatest" is raised, it always brings contention!

The question of "the greatest" is not a new question; the Lord's own disciples had a dispute about "the greatest" way back in Mark 9! **Who is the greatest in your life? Does the answer you would give match the answer that you live?** Who is number one in your life?

You can know who is greatest in your life by **how you treat children**. Kids can't inflate your ego, and they certainly can't repay you. They're kids! You can also know who is greatest in your life by **how you treat competitors.** The way you treat those who can not or will not benefit you is an indication of who is truly in first place in your life.

The Lord Jesus addressed both kids and competitors immediately after the disciples disputed about who would be the greatest. *"Whosoever shall receive one of such children in my name, receiveth me…. For he that is not against us is on our part."*

Do what you want with automobiles, computers, and football teams; but when it comes to the Lord Jesus, the only way life works is if Christ is first. Who is the greatest in your life?

◇◇◇

WHAT MOTIVATES YOU?

DAY 240

Mark 14:9 Verily I say unto you, Wheresoever this gospel shall be preached throughout the whole world, this also that she hath done shall be spoken of for a memorial of her.

In Mark 14:3-9 Mary spent a year's worth of wages in the form of ointment on the Lord Jesus. It was an extraordinary act that did not go unnoticed. Unfortunately, it was noticed by people who did not applaud her for her act of generosity, and, in fact, misunderstood it and even complained about it. Judas led the other disciples in characterizing what Mary had done as a waste. I think it's very instructive that Jesus, on the other hand, said that it was a good work and that what she had done should be and would be kept in memory for the ages.

What motivated Mary? First, we see her love. You wouldn't spend a year's salary on someone you do not love. The second thing that motivated her was her faith. The Bible says she did this looking towards the Lord's burial. The third motive was her overwhelming sense of gratitude. Yes, she loved the Lord Jesus and believed what he said; but this was an extraordinary act that was in response to an extraordinary gift she had been given. Namely, Jesus had raised her brother from the dead. This lady was immensely thankful for this extraordinary gift.

What about Judas? Despite all his pious talk, he complained under the guise of advocacy for the poor. He didn't care for the poor; he was a thief and held the bag. No one knew that except for Jesus. Here's a man who was a professional. He wasn't motivated by love for the Lord Jesus. He wasn't motivated by faith. Perhaps he expected Jesus to establish some great kingdom and make him the secretary of the treasury. And perhaps this very incident clarified to him that Jesus had different priorities than he did.

Now here's the takeaway. You cannot be motivated by greed and gratitude at the same time. Stop and think about all you've been given. We tend to be consumed by what we can get. **A person wastes his life if he is motivated by what he can get; but a person invests his life when he's motivated by what he has been given.** God has given us much. Let's be mindful and thankful, and let's invest what he has given to us to steward.

◇◇◇

DAY 241

THE ISSUE IS JESUS

Mark 15:3 And the chief priests accused him of many things: but he answered nothing.

Why was Jesus Christ crucified? Now we know that our sins put Jesus on the cross, and we know that His death, burial, and resurrection were all part of God's divine plan. But regarding those who judged Jesus that day, what was their reason for condemning Him?

Mark 14 tells us that false witnesses brought accusations against Him, and then *"the high priest stood up in the midst, and asked Jesus, saying, Answerest thou nothing? what is it which these witness against thee?"* The Bible says that Jesus "held his peace and answered nothing." Jesus did not respond to the accusations.

But His silence was broken when again the high priest asked him, *"Art thou the Christ, the Son of the Blessed? And Jesus said, I am....And they all condemned him to be guilty of death."* He did answer this question because this was the real issue.

Jesus will never be palatable to those who refuse, resent, or reject His authority and kingship. In short, Jesus is the issue, not something He demands, asks, or teaches. The issue is Jesus Himself. The rest of this chapter demonstrates this truth.

For instance, Mark 15:26 says, *"And the superscription of his accusation was written over, THE KING OF THE JEWS."* What was the accusation? Well, whether from a Roman or Jewish perspective, Jesus was crucified because of sinners' rejection of His place, His kingship, His lordship, His authority.

Jesus' rule is the real question. People reject Christ, not because of His commands or teachings, but because they would prefer to conform to rules rather than yield completely to Jesus' authority.

The issue is Jesus and whether He has authority in your life. If Jesus has complete authority in life, you will not quibble over what you should give, do, watch, hear, see, or say. **Recognize His lordship, and all these things will fall into place.**

DAY 242

LET JESUS SPEAK FOR HIMSELF

Mark 15:10,39 For he knew that the chief priests had delivered him for envy...And when the centurion, which stood over against him, saw that he so cried out, and gave up the ghost, he said, Truly this man was the Son of God.

Among all the people you told about the Lord Jesus in the last year, who was the most difficult to convince? Probably, this person was religious and knew something about the Lord Jesus. This person was hard to convince because knowing nothing about Jesus is better than believing the wrong things about Him.

When Jesus was brought to Pilate, He was brought by a group of very religious Christ-rejecters. The people thinking the most clearly were the pagan Romans. They seem to have had a grasp of what was happening more clearly than anyone else.

Let me give you two examples. Pilate represented Rome. He cared nothing about the Jews' religion. In turn, the religious leaders hated Pilate and wanted nothing to do with him. But the story they told Pilate was that Jesus was worthy of death because He was promoting insurrection against Rome. In actuality, nothing could have made them happier than an insurrection against Rome; but the story they told Pilate was that Jesus deserved to die because He was a rebel. The real issue was that the religious leaders were against God's Christ. Pilate sees this very clearly. Mark 15:10 says, *"For he knew that the chief priests had delivered him for envy."* Pilate was smart enough to see through their story and know the truth.

Another example of a clear thinking Roman is found in Mark 15:39. After Jesus gave up the ghost, the centurion (a pagan) says, *"Truly this man was the Son of God."* This illustrates that the more religious you are, the more difficult it will be to see Jesus as He is. A person can be religious and not be spiritual. Most people are religious, yet have rejected Christ. Believing something about Christ that is not true is worse than knowing nothing at all.

Be sure to let Christ speak for Himself. Is He a rebel, a revolutionary, a good man? Is He King of Kings and Lord of Lords? Is He the King of the Jews as He claimed to be? Let Christ speak for Himself; and if you have opportunity to speak for Him to others, just repeat what He has already said. At the end of the day the only thing that is eternally relevant is the truth. What Jesus has said about Himself is true. We need to take Him at His word and share that word with others.

EMPTY NETS

DAY 243

Luke 5:5 And Simon answering said unto him, Master, we have toiled all the night, and have taken nothing: nevertheless at thy word I will let down the net.

Michigan is a beautiful state with gorgeous lakes, but Michigan is also home to the stingiest fish in America. I know this firsthand! Earlier this year, my sons and I went fishing with a pastor friend; and while both of my boys and the pastor caught something, Yours Truly caught absolutely nothing. I am convinced the fish are stingy and don't like the thought of being hooked by some Tennessee preacher!

In Luke 5, Peter had been fishing all night and caught exactly nothing. I know that feeling! Peter caught nothing after a long night of fishing, but he caught a net full when he responded in obedience to Christ's command. What was the difference? His work the second time was in obedience to Christ. **Faith moves the God of heaven!**

In the work of the Lord, does it ever feel like you have toiled all night but nothing has happened? So few people today know what it means to work well and to work hard. When you see a hard worker these days, he stands out from the crowd.

But God is moved primarily by our faith, not by our work. That is true for salvation, but it is also true for living the Christian life every day. Often we think that if we can just work harder and longer at doing right, we will somehow gain the favor of our Heavenly Father. No, sir! God is pleased by our faith. **Faith in Christ today produces the very work that pleases God.**

Your hard work is not a virtue if it stands alone. Hard work is not wrong, but it is pointless if it is not the product of faith in God. God is not moved by our work so much as He is moved by our faith.

DAY 244
TRUST GOD, NOT MANIPULATION

Luke 6:31 And as ye would that men should do to you, do ye also to them likewise.

We're going to talk about a verse that you have preached. You say, "Brother Wil, I'm not a preacher." Well, you may not be, but I'm pretty sure that every adult looking at this has preached this before. Luke 6:31 says, *"And as ye would that men should do to you, do ye also to them likewise."* We sometimes call this the Golden Rule. There's hardly a teacher, parent, uncle, aunt, or camp counselor that has not taught this truth to some young person. The thought is that if you'll be nice to people, people will be nice to you.

It is true. Do unto others as you would have them do unto you. If I will do right by people, people will do right by me. The question is, "Why?" Is this merely psychological? Because of the way people are wired, they just can't help but respond in accordance with the way I treat them? Is this manipulation of people? Is this merely a way to get what I want from people?

If you look at Mark 6:32-34, you find that doing good to people merely because they've done good to you is nothing special. Jesus says that sinners do the same thing. It's natural, not supernatural. It's human, not divine.

Jesus says in verse 35, *"But love ye your enemies, and do good, and lend, hoping for nothing again."* Do you do that? The Bible says that your reward shall be great if you do these things. Verse 35 continues, *"Ye shall be the children of the Highest: for he is kind unto the unthankful and to the evil."* That is supernatural. When I, as God's child, follow His example, I live in God's power and trust Him to reward.

Now, God is the source of the reward, but people are very often the means. Verse 38 says, *"Give, and it shall be given unto you."* When I give, men will give unto me. God gives, but He doesn't usually sprinkle dollar bills from the sky. So what's happening here? God is the source; people are the means.

A man who trusts God will not feel the need to manipulate people. If I am trusting in God and enabled by God, I will do what He would do. I will treat others as I would want to be treated, and I will treat others as God has treated me. In short, I will do what is right and trust God to reward that in my life.

A DIFFERENT PERSPECTIVE

DAY 245

Luke 7:7 Wherefore neither thought I myself worthy to come unto thee: but say in a word, and my servant shall be healed.

There are three perspectives on your life today: the one you have, the one others have, and the one God has. Obviously, the one God has is the most important perspective, but the others aren't insignificant either.

In Luke 7 we find a fascinating story that illustrates these three perspectives. The Bible tells us the story of a centurion who was a Roman and a man of authority. He had a servant who was dear to him and who was ready to die because of an illness. The Bible says that he sent the elders of the Jews to ask Jesus to come and heal his servant. When these messengers came to Jesus, they asked Him to heal the servant and told Jesus that the centurion was worthy of this request.

Why did these messengers think that the centurion had worth? Well, he was a man of power, but he was also a man of grace. He wasn't asking help for his son, but for a servant. Here's a man powerful enough to have a servant, but gracious enough to care for the servant he had. He also loved the Jewish people and had helped build a synagogue for them.

Well, how did the centurion perceive himself? The Bible says that when Jesus was coming to the centurion's house, the centurion sent friends to tell Him, *"Lord, trouble not thyself: for I am not worthy that thou shouldest enter under my roof: Wherefore, neither thought I myself worthy to come unto thee."* So the centurion's perspective of himself was that "I'm not worthy, and I never thought I was worthy." His perspective was different than that of the messengers who presented his case to Christ.

What was Jesus' perspective of this man? When Jesus heard these things, He marveled. Can you imagine making God the Son marvel? Jesus said to the people that followed him, *"I say unto you, I have not found so great faith, no, not in Israel."* Even among the Jewish people, Jesus had not found this kind of faith. Perhaps this man had so much faith in Christ precisely because of the way he viewed himself. This proper perspective helped him view Jesus correctly.

There is a lesson to learn when we bring into focus these three perspectives. **People see what I do, but Jesus knows who I trust.** You will not impress God with what you do, but you will please God when you trust in Him. If you are trusting Him, you'll do exactly what you should do by His power.

THE LORD IS CREDIBLE

DAY 246

Luke 8:22 Now it came to pass on a certain day, that he went into a ship with his disciples: and he said unto them, Let us go over unto the other side of the lake. And they launched forth.

Throughout the thousands of years that men have told stories, they have generally divided these stories into three parts— the introduction, the conflict, and the resolution. I think this is true of the story we see in Luke 8.

Luke says of Jesus, *"Now… he went into a ship with his disciples: and he said unto them, Let us go over unto the other side of the lake. And they launched forth."* As soon as they set off, Jesus fell asleep. Subsequently, there was a storm of such magnitude that even some of the disciples who were seasoned fishermen were in fear for their lives. So they woke the Master and said, *"We perish!"* I think the disciples were surprised that He was sleeping, and He may have been disappointed that they were surprised.

"Then he arose, and rebuked the wind and the raging of the water: and they ceased, and there was a calm." The calm that was in Jesus as He slept was the calm that now characterized the waters that had just been in such turmoil. *"And he said unto them, Where is your faith? And they being afraid wondered… What manner of man is this! for he commandeth even the winds and water, and they obey him."*

This story has three parts: a storm, a command, and the peace of Jesus. I don't know where you are in your story. Maybe you are in the introduction and need to pay attention to what the Lord is telling you. Maybe you are in the middle, where the storm is so ferocious that you feel you will not make it. Maybe you are on the other side of the storm looking back, and you have forgotten just how grateful you should be. **The Lord who commands the wind and waves has the credibility to lead you. There is a calm that comes when you follow His lead.**

DAY 247

MISS JESUS, MISS THE POINT

Luke 9:35,36 And there came a voice out of the cloud, saying, This is my beloved Son: hear him. And when the voice was past, Jesus was found alone....

God is very much in earnest about highlighting the singular nature of His Son, Jesus Christ. Jesus is unique. He alone occupies His position. He is singular. There is none like Him. You find that repeatedly emphasized in the stories of Luke 9. For example, in Luke 9:7 Herod was perplexed about Jesus. Some people said He was John the Baptist raised from the dead. Others said He was Elijah or one of the old prophets. No, he was not one of these men. He was Jesus.

Again, Jesus asked His own disciples who people thought that He was. They said that some people said He was John the Baptist, and some said Elijah or one of the old prophets. People found it easier to believe that an old prophet had come back from the dead than to acknowledge that Jesus was God in flesh, the Messiah, and God's Son.

Beginning in Luke 9:28, the Bible tells us the story of Jesus taking Peter, James, and John to the Mount of Transfiguration. In that instance two prophets, Moses and Elijah, came back from the dead. Peter saw this and thought, "Whoa, this is amazing. Here are Moses, Elijah, and Jesus, and these are all great men." It seems that Peter almost saw Jesus on the same level as just another great prophet. But God answers dogmatically, *"This is my beloved Son: hear him. And when the voice was past, Jesus was found alone."* They saw Jesus only.

If you miss Jesus in the events of this day, you've missed the point. It is so easy to be in awe of other people or in awe of what God has done but miss Jesus Himself in the works that He has done. Sometimes people are the means of God's working, but never forget that the Source of the wonderful works of God is God Himself. The Lord Jesus is singular. He will not share a place with anyone else in your life. He deserves our highest praise and recognition in the events of this day.

WHERE IS YOUR FOCUS?

DAY 248

Luke 10:41-42 And Jesus answered and said unto her, Martha, Martha, thou art careful and troubled about many things: But one thing is needful: and Mary hath chosen that good part, which shall not be taken away from her.

Sometimes it's the question at the end of the day that discourages me the most. Maybe I've gotten less sleep than I'd like, run a little bit late in the morning, and then worked hard all day…one thing after another. I drag myself back home, and someone asks the question, "What did you do today?" I think to myself, "I don't know." There have been many days that I've worked hard, but I honestly don't know what I can show for it. I don't want to make that an everyday occurrence. But I can do my work more effectively today if I remember why I'm doing it and for Whom I am doing it.

In Luke 10 we read the story of Jesus' visit to the sisters, Mary and Martha. The Bible says that Martha received Him into her house, and what follows is based on this basic premise that Martha felt responsible for the guest in her house. There's nothing wrong with that. So, we don't need to be too hard on Martha. The point of the story that follows is not that you need to just sit and meditate all day and let other people do the work. The point of the story is the importance of focus in our lives.

Mary sat at Jesus' feet and heard His word, but Martha was cumbered and distracted. She was distracted from her guest, her Lord Jesus. The Bible says that Martha was distracted with much serving. Now, serving Jesus wasn't bad. But, the Bible says she came to Him and said, *"Lord, dost thou not care that my sister hath left me to serve alone? bid her therefore that she help me."* But Jesus answered, *"Martha, thou art careful and troubled about many things."* She was focused on what she was doing, but not on Whom she was serving. In verse 42 Jesus says to Martha,

"But one thing is needful: and Mary hath chosen that good part, which shall not be taken away from her."

Serving is important, and if you're not serving I would say it's not because you're thinking about the Lord Jesus too much! The point is not that we should stop serving, but that **serving should be our response to Jesus and not a distraction from Him.**

◇◇◇

DAY 249 — SERVING CHRIST ENRICHES ME

Luke 17:10 So likewise ye, when ye shall have done all those things which are commanded you, say, We are unprofitable servants: we have done that which was our duty to do.

I've heard it said a number of times recently that it's easy to be a servant until someone treats you like you are a servant. It's almost a Christian honorary title to be called a servant, but there's a difference between using "servant" as a benevolent title that shows appreciation and actually doing what a servant does.

Luke 17: 7 says, *"But which of you, having a servant plowing or feeding cattle, will say unto him by and by, when he is come from the field, Go and sit down to meat?"* No, that doesn't happen. Verse 8 says *"And will not rather say unto him, Make ready wherewith I may sup, and gird thyself, and serve me, till I have eaten and drunken; and afterward thou shalt eat and drink? Doth he thank that servant because he did the things that were commanded him? I trow not. So likewise ye, when ye shall have done all those things which are commanded you, say, We are unprofitable servants: we have done that which was our duty to do."*

Is Jesus suggesting that His servants are worthless? No. What He's saying it that when I serve Christ, He doesn't owe me anything in return. Jesus is not getting rich off of our work. When we serve Christ we are doing that which is our debt or duty to do.

Serving Christ does not enrich Him; it enriches me. Earlier in the chapter, Jesus told the disciples that if a man *"trespass against thee seven times in a day, and seven times in a day turn* *again to thee, saying, I repent: thou shalt forgive him."* Then Jesus talks about faith and launches into this discourse about servants.

So, here is a guy who does me wrong over and over again; but when he asks, I forgive. Someone says, "Wow, what an amazing servant of God this Wil Rice is to forgive the same person so many times." Christ isn't being enriched by that; I am.

In the first place, if Christ tells me to forgive this way, who do you think already forgives that way? Christ does. Whom has Christ forgiven that way, and whom does He continue to forgive that way? The answer is, "Me."

Second of all, if I am forgiving like Christ would, who is actually supplying the power to forgive as Christ would in my life? Again, the answer is Jesus. I'm doing what I'm a debtor to do, and this doesn't profit the Lord Jesus. It profits me.

◇◇◇

GOD'S PURPOSE FOR ME

DAY 250

Luke 19:31 And if any man ask you, Why do ye loose him? Thus shall ye say unto him, Because the Lord hath need of him.

If you're anything like me, you like to attach as much significance and purpose to everything you do as you possibly can. Sometimes that's difficult because most of our days are defined by tasks like preparing meals, preparing a report, or finishing homework. Sometimes it's hard to see God's grand scheme in what I'm doing right now. We have the need to recognize a grand purpose and a great God at work in the world. But what if there are things that must be done or that God asks of us which are hard to attach to some greater significance?

I think this scenario has been illustrated for us in Luke 19. Jesus sent a couple of His disciples into the village ahead of Him and said, *"Ye shall find a colt tied, whereon yet never a man sat: loose him, and bring him hither."* Now, if you'd been one of Jesus' disciples, what would be your first question? Wouldn't you at least wonder why?

Verse 31 says, *"And if any man ask you, Why do ye loose him? thus shall ye say unto him, Because the Lord hath need of him."* The disciples didn't know what was going on here, and I don't know what the owner of this colt must have thought. The Bible tells us that the disciples did just as Jesus said. They were loosing the colt, and the owners said to them, *"Why loose ye the colt?"* And they said, *"The Lord hath need of him."*

If you read the rest of the chapter, you get the sense that there was something grand going on. Thousands of people were involved, but I don't know how many of them actually understood their part. There were some that praised the Lord, and some that sneered at Him. There was on who had a colt and gave it away, and some that took the colt for the Lord Jesus. I'm not sure any of them comprehended God's grand plan for history, but there was one.

There's a big difference between having no purpose and having no understanding of what that purpose is. It's possible to be involved in a greater purpose but not comprehend it. Now I would rather be mindful of what God is doing in any given moment or task, but what about those times when there are things that must be done simply because I know that it's the right thing right now?

Those are the times that I need to trust and obey God even though I cannot see the grand purpose. I don't need to understand God's grand purpose for the way He's leading me in order to know that there is one. Trust God. Do what you should. And if you do, even the small issues of life may have far grander significance than you could possibly know this day.

DO YOU REALLY WANT AN ANSWER?

DAY 251

Luke 20:2 And spake unto him, saying, Tell us, by what authority doest thou these things? Or who is he that gave thee this authority?

If you're a parent, you probably have had a child who asked the same question incessantly until he got the answer he wanted. There's a difference between asking a question for *an* answer and asking a question for *the* answer. Adults also ask questions, not necessarily wanting an answer but wanting something else, like an argument. There are no answers for those who just want arguments. Argumentative questions are what Jesus received from the religious leaders in Luke 20.

Luke 20:2 is probably the best question they had, and it's also the question that summarizes the essence of all the questions that follow. *"They spake unto him, saying, Tell us, by what authority doest thou these things? Or who is he that gave thee this authority?"* Now that's a valid question. It's just that those who are most concerned about their own authority are often the least submitted to God's. They felt threatened by Jesus because they were not in submission to the Father who had sent Him.

In verse 22 they said, *"Is it lawful for us to give tribute unto Caesar, or no?"* Once again this question was valid, but the motive was argument. Oftentimes, instead of asking "How can I obey?" our question is "Against whom can I rebel?" Rebels are hypersensitive to conflict because they prefer conflict over obedience.

There are no answers for those who want arguments. These men eventually stopped asking questions because they weren't getting the arguments they wanted to use as fodder against Jesus.

Submitting to Jesus Christ answers questions that nothing else can. There are some questions that cannot be answered by intellect or by argument, only by submission. When we submit to Christ, then suddenly life makes more sense. God never denies an answer to an open hand, an open heart, and an open mind; but He resists the proud.

DAY 252
YOUR RESPONSIBILITY

Luke 23:7 And as soon as he knew that he belonged unto Herod's jurisdiction, he sent him to Herod, who himself also was at Jerusalem at that time.

Will Rice, my great-grandfather, was saved in 1887 in a little Baptist church in Texas. He was a cowboy, a state senator, and a preacher. Three of his sons became preachers, including my grandfather, Bill Rice. My dad is a preacher, and I am one, too! Do you think God is impressed with my preacher heritage? Do you think that God will cut me some slack because my great-grandfather was a preacher? No sir, not on your life!

My heritage is not insignificant; it is significant precisely because I am responsible for what I do with my heritage and what I do with the Lord Jesus. I am thankful for my Christian heritage and for the preachers in my family who have come before me, but I would be foolish to use either fact to evaluate my relationship to God.

Your relationship to Christ is a personal matter that is *your* responsibility, and you will be judged for how *you* judge Christ. In Luke 23, Pilate clearly said, *"I find no fault in this man."* But he did not act on what he knew because he wanted to pass responsibility to someone else. The Bible says, *"And as soon as he knew he belonged unto Herod's jurisdiction, he sent him to Herod...."* Pilate found no fault in Christ, but he did not want to be responsible for his response to Christ personally.

All of us are responsible for how we judge the Lord, how we follow the Lord, and how we know the Lord. God will judge me for how I have judged the Lord Jesus. Your relationship with the Lord Jesus is *your* responsibility.

YOUR EXPECTATION OF JESUS

DAY 253

Luke 24:21 But we trusted that it had been he which should have redeemed Israel: and beside all this, to day is the third day since these things were done.

One of the most fulfilling things in life is getting to know people better. This also happens to be one of the most disappointing things in life. Haven't you met someone you thought very well of and then, as you got to know them better, were disappointed because they didn't quite measure up to your expectations? I've come to the conclusion that many times it is not that people have changed. My perception of them has changed as I've gotten to know them better.

In any event, the disciples of the Lord Jesus were greatly discouraged. He had died, and they had seen it. A couple of these disciples were on a road to a village called Emmaus. They were talking about all that had happened—how they had expected Jesus to do one thing, and something else had happened. Jesus, the risen Christ, joined them on this trip and said, "Why are you sad?" They said, "Don't you know what has happened? We had great, high hopes for this Jesus of Nazareth. And now He has died."

Jesus did not measure up to their expectations. As it turned out, He was far greater than their expectations. His stern reply came in verse 25, *"O fools, and slow of heart to believe all that the prophets have spoken: ought not Christ to have suffered these things, and to enter into his glory? and beginning at Moses and all the prophets, he expounded unto them in all the scriptures the things concerning himself."* The Jesus revealed in the Bible is so much better than the Jesus that we can imagine on our own.

It may be that you're disappointed that Jesus has not met your limited expectations. Jesus doesn't live down to our expectations; He lives up to His Word. That's why it's so important to know what God has said about the Lord Jesus. The more you know God's Word, the better you know God's Son.

There are two things you need in order to know who Christ is. First, you need to open His Word. Secondly, you need Him to open your understanding. Without an open Bible and an open understanding, our expectations are not what they ought to be. Jesus will not live down to your expectations; He will live up to His Word. That means that your level of faith and confidence comes from your level of understanding of what God has said.

Know what God has said, open God's Word, and allow God to open your understanding in order for your expectations to rise to the level of Who Jesus really is.

◇◇◇

DAY 254
WORD OF MOUTH

John 1:37 And the two disciples heard him speak, and they followed Jesus.

Think back to the last two books you've read and the last two major purchases you've made. Both probably trace back to people who influenced you. The truth is, **we win others over to the people and things that we see as important.**

Two of John's disciples were with him one day when John saw the Lord Jesus and said, "Behold the Lamb of God!" He directed their attention to the Lord and promptly lost two of his own disciples. John 1:37 says that the disciples heard John speak and followed the Lord. John literally caused himself to lose two disciples! Was this a loss or a gain?

If John saw himself as important, this would've been disastrous. Imagine how different that day would have been if John saw the Lord as a threat to his work! Instead, John's priority was to be the forerunner for the Lamb of God. He didn't lose two disciples; he accomplished the very thing that was most important!

People are moved by other people. You will be confident and buy into whatever the people you trust buy into. For good or bad, you will win people to whatever and whomever you see as important. Word of mouth still matters.

After John's two disciples followed Jesus, you see the pattern of winning people by word of mouth continuing. Andrew, one of the two disciples, found his brother Peter and said, *"We have found ... the Christ."* The next day, Philip followed Jesus and found Nathanael and said, *"We have found him ... Jesus of Nazareth."* This all started because John directed two of his own disciples' attention to the Lamb of God!

To what or to whom are you winning people? You will win others to the people or things you see as important. Are you winning people to yourself and to your own ambitions? Or are you like John, winning people to Jesus Christ regardless of the loss to yourself?

AS SIMPLE AS ONE

DAY 255

John 3:18 He that believeth on him is not condemned: but he that believeth not is condemned already, because he hath not believed in the name of the only begotten Son of God.

Maybe you've heard it said that the gospel is as simple as A,B,C and 1,2,3. That is certainly true. In fact, I would say that salvation is as simple as One. Salvation is not in a plan; it's in a Person. It's not in a way; it is in *the* Way which, again, is a Person, the Lord Jesus.

In John 3, Jesus tells us over and again what is required in order to have peace with God and to have eternal life. In John 3:15 He says, *"That whosoever believeth in him should not perish, but have eternal life."* So how does one not perish and have eternal life? By believing, depending on, trusting in Jesus Christ. Verse 16 says, *"For God so loved the world, that he gave his only begotten Son, that whosoever believeth in him should not perish, but have everlasting life."* So, everlasting life comes from a Person, Jesus Christ.

Verse 18 says, *"He that believeth on him is not condemned: but he that believeth not is condemned already, because he hath not believed in the name of the only begotten Son of God."* Verse 36 says, *"He that believeth on the Son hath everlasting life: and he that believeth not the Son shall not see life; but the wrath of God abideth on him."*

In II Corinthians 11:3, the Bible speaks of the simplicity of the gospel as opposed to the subtlety of falsehood. The word *simplicity* there means "singular" or "one." There are not twenty-seven steps to salvation. Salvation is as simple as One. Well, what must a person do to be saved? Many answer, "Repent, receive, accept, trust, or believe." Yes, but these are not five or six different ways of salvation. It may be that there are several different ways to express the way to salvation; but there is only one way, and that way is a Person, Jesus Christ.

Only one sin can damn a person, and that sin is not believing on Jesus. He that believeth not is condemned. When? Already. Why? Because he has not believed in the name of the only begotten Son of God. Only one thing saves: believe on the Lord Jesus Christ.

◇◇◇

DAY 256
YOUR DEFINING RELATIONSHIP

John 8:53 Art thou greater than our father Abraham, which is dead? and the prophets are dead: whom makest thou thyself?

I love history. I think we, as a culture, are very short-sighted. We rarely think about what life was like one day before we were born or will be like one day after we are gone. Most of us have no context for our lives or the times in which we are living. But today we are living in an extraordinary, "watershed" time in history.

With that being said, most of us end up defining our very identity by who we are and not who God is. Jews who were proud of their heritage asked the God of Abraham, *"Art thou greater than our father Abraham? … whom makest thou thyself?"* You may not be Jewish, but you might have a godly heritage. I love my family, and I am thankful for the good heritage I've been left. But if I am not careful, I can define my life by who my grandfather *was* instead of who God *is*.

Your relationship to God Almighty is the defining relationship of your life. If you are right with God, you will be right with your spouse, children, and parents. You can't be right with the God that you cannot see if you are not right with the people you can see (I John 4:20). Thank God for the good heritage you enjoy, but don't let the piety of your heritage dim the truth that the defining relationship of your life is your relationship to God and His Son.

GOD DWELLS WITHIN

DAY 257

John 12:11 Because that by reason of him many of the Jews went away, and believed on Jesus.

Have you ever had guests come to your home when you were a parent with young children? Kids can totally change the dynamic of the meal! Any parent in this situation thinks of a lengthy list of "don'ts" just to ensure that everyone is on their best behavior. Now, can you imagine having the *Lord Jesus* come to your home? Mary, Martha, and Lazarus did in John 12!

Martha felt responsible for the meal and hospitality, and the Bible says that she served. As for Lazarus, he sat down with the very One who had raised him from the dead. What about Mary? She brought an expensive ointment and anointed the feet of the Lord Jesus. In fact, a crowd of people heard about

Martha's guest and came to see both Jesus and the man He had raised from the dead, Lazarus.

It is not surprising that the religious leaders of the day were not keen on the popularity and attention Lazarus began receiving. The chief priests *"consulted that they might put Lazarus also to death; because that by reason of him many of the Jews went away, and believed on Jesus."* What a reputation and testimony!

If God has done something in your home, there ought to be thanksgiving that is visible to other people. When God has worked in your life, others should see and know what He has done. **It is hard to hide a house where God is!**

When we look at the wicked world around us, our tendency is to "hold the fort" instead of "storm the gates." In our homes and churches, it is not God's intent that we sit back and bunker down. **His intent is that our families affect our world.**

Has the Lord made a difference in your home? Is your testimony at work or school drawing others to Christ? Are people coming to the Lord because of you? People in Lazarus' day couldn't ignore what God had done in his life. Let your home shine brightly in this dark world today!

DAY 258
A SERVANT IS NOT GREATER THAN HIS LORD

John 13:16 Verily, verily, I say unto you, The servant is not greater than his lord; neither he that is sent greater than he that sent him.

There is no one greater than Jesus Christ, and there is no greater servant than Jesus Christ. Thus our relationship to all other people is a reflection of our relationship to the Master, this Servant among men, the Lord Jesus Christ.

Jesus reminded His disciples in John 13:16 that the servant is not greater than his lord. He said this after a very personal act of service, washing the disciples' feet. It was an act of great humility by God in human flesh, and He said, *"I have given you an example, that ye should do as I have done to you."*

That **the servant is not greater than his lord** is an important truth that affects every area of our lives. Are there people who have nicer houses than you have? Well, of course. Did Jesus have a nicer house than you have? The servant is not greater than his lord. Are there people that have more education, more titles, or more degrees than you have? Certainly. Does Jesus have more? Was Jesus more highly regarded by the people of His day than you are by the people of yours? No, the people of Christ's day crucified Him. Are you ever misunderstood? Well, of course. Do you think that the Lord Jesus was misunderstood any less than you have been? The servant is not greater than his lord.

There are three settings that specifically drive this home. Verse 16 says, *"The servant is not greater than his lord; neither he that is sent greater than he that sent him."* Would He serve those that you would not? The servant is not greater than his lord.

In verse 20 He says, *"Verily, verily, I say unto you, he that receiveth whomsoever I send, receiveth me; and he that receiveth me receiveth him that sent me."* Are there those whom Jesus has sent that you will not receive? Are there those whom Jesus would receive that you would not receive?

What about love? In verse 35 Jesus said, *"By this shall all men know that ye are my disciples, if ye have love one to another."* Love is not a feeling alone. It is not something as fickle as "like." No, love is doing what is right by people whether I like them or not. It is a decision. It is Christ's power through me to love them as Jesus loves me and others. So, would Jesus love those that you would not?

These are questions that we should ask ourselves every day throughout the day. We would do well to remember that our relationship to others is but a reflection of our relationship to the Master.

DAY 259

WHAT JESUS WOULD DO IS WHAT I SHOULD DO

John 16:14 He shall glorify me: for he shall receive of mine, and shall shew it unto you.

What can you do? Can you run a sub-5 minute mile? Can you speak three different languages? Have you read *War and Peace*? There may be many things you can do, but there are certainly many things that neither of us can do. What can Jesus do? Someone says, "Anything. Everything." And, of course, that is true. The more compelling question is "What *would* Jesus do?" Someone says, "Anything." Well, no, of course not. Jesus can do anything, but He only wants to do what is the Father's will to do.

In John 16 Jesus talks about the Comforter, the Holy Spirit. He says of the Holy Spirit in verse 14, *"He shall glorify me."* The Holy Spirit can do what I cannot do by natural inclination or by my natural power. He can glorify the Lord Jesus and the Father.

Jesus owns many things that I do not. If you look at John 15:8-11, Jesus speaks of "my disciples," "my love," "my commandments," and "my joy." In John 15:26-27, Jesus talks about "my message," and He says the Holy Spirit "shall testify of me," and "ye also shall bear witness." So Jesus basically says, "Look, the Holy Spirit of God is going to tell people about Me, and you are going to tell people about Me as my disciples." When does the Holy Spirit tell people about Jesus? The answer is, largely and often, when I do. The Spirit tells them through me.

How can I obey Jesus' commandments? How can I have Jesus' love? How can I have Jesus' joy? The answer is by being His disciple. When Jesus owns me, I have everything I need. This would include Jesus' joy, Jesus' commands, Jesus' love, Jesus' message, and Jesus' power. Philippians 2:13 says, *"For it is God which worketh in you."* What does God work in me? He works in us *"both to will [to desire] and to do [to perform] of his good pleasure."*

What Jesus would do is what I should do. If you will let the Holy Spirit of God have His way in your life today, you can ask yourself the questions, "What would Jesus think? What would Jesus say? What would Jesus do?" Better yet, let the Spirit of Christ do what needs to be done through you. The Holy Spirit of God will guide you to the Word that Jesus has said, help you see what Jesus has done, and give you the heart to do what only He can do.

ARE YOU A COWARD?

DAY 260

John 18:21 Why askest thou me? ask them which heard me, what I have said unto them: behold, they know what I said.

Would you like to know some top-secret, highly classified piece of information? Well, did you know that the truths of Scripture are not top secret or classified? When the Lord Jesus was questioned about His doctrine by the high priest in John 18, He answered by saying, *"I spake openly to the world … in secret have I said nothing … ask them which heard me … they know what I said."* Where were the people whom Jesus had taught and mentioned specifically? Were they by His side? No sir! They were hiding!

Cowardice is the gap between what you know and where you stand. Knowledge costs you nothing; acknowledging the truth does. Bible-believing Christians have caved in to the world's ideas about marriage, the home, and God's truth. If you look to the world or even to another person to affirm what you believe, you believe nothing.

Where do you stand? You may *say* one thing, but it matters what you *do*. Judas Iscariot *"stood with them,"* the enemies of Christ.

Peter *"stood at the door without."* Both were cowards at that time when it came to acknowledging the truth about Jesus Christ.

I have to admit, I have found myself wishing that the world could just leave us Christians alone. That is passive and cowardly, my friend! The problem is that we are leaving the world alone. Just as the disciples in John 18, those who know the Lord Jesus the best are speaking for Him the least when it actually matters. May God help us to stand for what is right, say what is right, and follow what is right.

DAY 261
THE SOURCE OF ALL AUTHORITY

John 19:11 Jesus answered, Thou couldest have no power at all against me, except it were given thee from above: therefore he that delivered me unto thee hath the greater sin.

Isn't it amazing how kids can work one authority against another, as in pitting Mom against Dad? That's human nature! Oftentimes, adults can act in much the same way. In John 19, we find a contrast between Christ's submission and the rebellion of the civil and religious authorities. Caesar, Pilate, and Caiaphas all had authority, but their authority stemmed from only one authority. That one authority was embodied in the man Jesus Christ.

The Lord made it very clear in verse 11 that the authority Pilate possessed was given to him by God. *"Thou couldest have no power at all against me, except it were given thee from above."* We are wise to remember **that all authority comes from heaven**. Romans 13:1 says it this way: *"For there is no power but of God: the powers that be are ordained of God."* There is one authority in this world, and that is God. He has delegated authority in many different ways to many different people. Not all authorities honor God or do right, but any authority a person has comes from the merciful hand of God.

Since all authority comes from heaven, **all authorities are accountable to heaven**. That means you stand accountable for what you do with the authority God has entrusted to you. Pilate tried to dismiss his responsibility by saying, "I find no fault in him," but that doesn't pass in the court of heaven. Don't pass the buck with the authority you have been given by God!

How are you stewarding the authority you have been given? How are you responding to the authority of others? If God has placed authorities in your life, it is good, healthy, and wise to voluntarily place yourself under that authority. Authority comes *from* heaven, and authorities are accountable *to* heaven.

WHO DO YOU KNOW?

DAY 262

Acts 4:13 Now when they saw the boldness of Peter and John, and perceived that they were unlearned and ignorant men, they marvelled: and they took knowledge of them, that they had been with Jesus.

Do you know a name dropper, someone who brags, not because of what they have accomplished, but because of who they know? We generally tend to dismiss such people and to look down upon them a little bit. I understand this, but when you read the book of Acts you realize that the early church was constantly name dropping. In fact, they weren't just dropping the name of Jesus Christ, they were broadcasting it.

The Bible tells us that Peter and John went into the Temple where they healed a lame man by the power of Jesus Christ. The religious leaders who had just crucified the Lord Jesus roughed up Peter and John a little bit. Then they asked a dangerous question, *"By what name have ye done this?"* Well, that sure was the wrong question to ask these men because every chance they got, they were proclaiming the name, the Person, the power, and the work of the Lord Jesus Christ.

In verses 17 and 18, the Bible says that these leaders panicked and determined that the preaching of Christ must not spread any further among the people. They decided to threaten Peter and John and command them not to teach or speak in Jesus' name. Their opposition was in response to Jesus. The message Peter and John had was because of Jesus. The power they had was because of Jesus.

Peter and John were not prosecuting attorneys making arguments, nor were they defense attorneys defending the Lord Jesus. They were witnesses, merely telling what they knew, telling Who they knew. That is where their power to heal this man came from. The power to change the world, and to make a difference. The power that we find in the book of Acts came by the acts of the Spirit of God, the Spirit which Jesus Christ Himself sent and left for us. Your power doesn't come from what you do. It doesn't come from what you know. **Your power comes from Who you know, the Lord Jesus.**

◇◇◇

DAY 263 — THE SPIRIT OF UNITY

Acts 5:12 And by the hands of the apostles were many signs and wonders wrought among the people; (and they were all with one accord in Solomon's porch.)

The book of Acts is a book of actions, and behind every action is a reason. One reason for God's actions is found in Acts 5:12 where it says, *"By the hands of the apostles were many signs and wonders wrought among the people; (and they were all with one accord in Solomon's porch.)"* Now this may not be what you consider a powerful verse, but even a verse like this carries extraordinary weight because of the other verses that stand upon its shoulders.

Right at the very start in Acts 1:14 the Bible says, *"These all continued with one accord in prayer and supplication."* Acts 2:1 says, *"They were all with one accord in one place."* The word *accord* comes from two different words meaning "to rush along" and "in unison." So the idea of being in one accord is "to rush along in unison." The people we read of here were uniformly submitted to one Person with one message.

As the book of Acts unfolds, you find that there were among the believers those who were rich and those who were poor, those who were Jews and those who were Gentiles, and those who were ladies and those who were gentlemen. But all of these had one mind and were of one accord. Acts 4:32 says again, *"And the multitude of them that believed were of one heart and of one soul."* Now about what were these people of one heart and one soul? It was the gospel of Jesus Christ.

The way to enjoy a spirit of unity is to be in submission to the Spirit of God. We should be just like a symphony that plays beautiful music. Every person has a different part to play, but is under the direction of and in submission to one conductor.

God is the One who has created diversity. He has gifted us in a variety of ways and given us a variety of strengths, but that is not what will make a difference. The difference is made when all of us from different places and with different strengths submit ourselves to Jesus Christ and allow His Spirit to have His way in our lives.

THE FOUNDATIONAL RELATIONSHIP

DAY 264

Acts 11:17 Forasmuch then as God gave them the like gift as he did unto us, who believed on the Lord Jesus Christ; what was I, that I could withstand God?

Doesn't it feel good to belong to something? All of us belong to something. You belong to a family, perhaps, to a church, and, in some sense, to a nation or country. All these things are important, and I give thanks to God for each of these in turn. But none of these "memberships" identify me to God. The sole relationship that identifies me to God is the relationship I have with His Son.

God taught Peter this lesson, and Peter recounts this lesson to his own countrymen who were contending with him because he was fellowshipping with people who loved Christ but were not Jewish. Peter says in Acts 11:17-18, *"Forasmuch then as God gave them the like gift as he did unto us, who believed on the Lord Jesus Christ; what was I, that I could withstand God? When they heard these things, they held their peace, and glorified God, saying, Then hath God also to the Gentiles granted repentance unto life."*

The only relationship you have that impresses God is the one you have with Jesus Christ. The Bible tells us that in Christ, we are accepted in the beloved. It is not my merit, title, or nationality that impresses God; it is my relationship to the Lord Jesus.

Now the other secondary areas of identification in my life have their place of importance, as well. It just so happens that the people with whom I enjoy the closest fellowship tend to share other things in common, but I must always remember that the foundational relationship that forms my identity is my relationship with the Lord Jesus.

Is it possible for someone to belong to my church but not belong to Jesus? Is it possible for someone to belong to my family but not to Jesus? Is it possible for someone to belong to my country but not to Jesus? The answers are yes, yes, and yes. The only relationship that I have that impresses God is the one I have with Jesus Christ. Whatever else you do today, make sure that relationship is what is ought to be. If it is, your other relationships will be too.

WHAT DO YOU EXPECT?

DAY 265

Acts 12:11 And when Peter was come to himself, he said, Now I know of a surety, that the Lord hath sent his angel, and hath delivered me out of the hand of Herod, and from all the expectation of the people of the Jews.

Isn't it interesting how our responses to life are governed by our expectations for life? That's why some people go through life expecting all of the worst things to happen to them, and they do. Other people can be walking around on the dreariest of days but have a certain shine about them because of what they expect in the long run.

Acts 12 is a wonderful example of how our responses are governed by our expectations. Peter had been taken prisoner by Herod. Herod had previously killed James, and because it was so popular, he intended to do the same thing with Peter. Well, God answered prayer and miraculously spared Peter. It's fascinating to see the expectations of other people and how they responded to Peter's release from prison.

For instance, Acts 12:11 says, *"And when Peter was come to himself, he said, Now I know of a surety, that the Lord hath sent his angel, and hath delivered me out of the hand of Herod, and from all the expectation of the people of the Jews."* What did all the people expect? They expected Peter to be killed, but he wasn't because God's plan exceeded their expectation.

When Peter was freed, he went to a house where there was a prayer meeting. It very well may be that the prayer meeting was on his behalf. The Bible says that when Peter came to the courtyard where believers were praying, they were astonished at his appearance. They could not believe that it was actually Peter, despite the fact that they had been praying that God would do just exactly what God did!

What did Peter expect? Well, the Bible tells us back at the very beginning of the story that Peter was sleeping; he wasn't worried because he expected to be spared. Why did he expect that? He was resting in the promise of God.

John 21 records that Jesus had told Peter how Peter would die, at least in a general sense. He told Peter that his death would happen when he was old. Peter's confidence came from what Jesus had said.

Is there always an explicit promise that you can apply to every situation of your day? No. There are times when you simply rest in the providence of God. That is, you know that God is in control, that God is good, and that He will do right regardless of what you can see. But there are other times when you have something even more than that. You have the Word of God, the promise of God.

Let your expectations be governed by what God says, not by what you see. Sometimes we don't see what really is. What we see can be flawed, but what God says never is. You can rest in His Word today.

◇◇◇

DAY 266
SHAKE IT OFF!

Acts 13:51 But they shook off the dust of their feet against them, and came unto Iconium.

Doing right has always drawn the opposition of those who are doing wrong. Being discouraged about doing what is right is nothing new! The Apostles were busy declaring the Good News, but not everyone was glad or in a good mood about it. They were even interrupted by a sorcerer named Elymas and eventually expelled from the country. Talk about a rough time on visitation outreach!

Have you ever had a terrible, lousy, all-together bad day while serving Christ? Those days have existed since the beginning. **If you do what is right, you will be opposed by the wrong.** Doing right doesn't mean that everything will be easy, wonderful, and uneventful. Like the Apostles, you may face opposition you would not have faced apart from doing right.

So what do you do in such cases? **Shake it off and keep it up!** Paul and Barnabas *"shook off the dust of their feet against them, and came unto Iconium. And the disciples were filled with joy and with the Holy Ghost."* The Bible is not giving an excuse for blowing people off or an alibi for giving up. It is encouraging us to keep going with the gospel message. As a result, *"a great multitude both of Jews and also of the Greeks believed."* That beats sitting around and sulking about the rude reception in Antioch!

There may come a time when it is better to "shake it off" and move on. Don't languish in defeat, discouragement, and opposition. It is a new day with new opportunities to spread the Good News. **You can't move on if you are stuck in one place.** Shake it off, and then keep on doing right. You don't have a choice about facing opposition, but you do have a choice in how you respond. You may have brand new opportunities waiting for you, so shake it off and keep at it!

◇◇◇

ARE YOU SUCCESSFUL IN GOD'S SIGHT?

DAY 267

Acts 14:18-19 And with these sayings scarce restrained they the people, that they had not done sacrifice unto them. And there came thither certain Jews from Antioch and Iconium, who persuaded the people, and, having stoned Paul, drew him out of the city, supposing he had been dead.

It is amazing how many people do not know how to drive stick shift these days. Have you ever heard someone grind the gears? It kind of makes you cringe, doesn't it? While reading Acts 14, I could almost hear the gears grinding when I read verses 18 and 19 about the work of the Paul and Barnabas. It's almost as if the Bible shifts from fifth gear, highway speed, directly into reverse as it talks about how people responded to the message these men gave.

Verse 18 says, *"And with these sayings scarce restrained they the people, that they had not done sacrifice unto them."* Earlier in verse 11 they had said of these men, *"The gods are come down to us in the likeness of men."* That seems like a pretty good day for Paul and Barnabas, if "good" was defined by how well people treated them. What these people were saying about Paul and Barnabas was blasphemous, but they obviously thought very highly of these men.

Verse 19 says, *"And there came thither certain Jews from Antioch and Iconium, who persuaded the people, and, having stoned Paul, drew him out of the city, supposing he had been dead."* In one moment, the Bible speaks of people wanting to treat these servants of God as gods, and in the next verse they want to kill them.

If you are looking to others to gauge your success today, it is going to be a roller coaster ride. Some people will praise you, and others may want to throw rocks at you. A wise person doesn't listen too much to the praise that he's given, nor does he get too discouraged by the people who have it out for him.

Success is determined by how you respond to God, not by how people respond to you when you do. On one hand people loved Paul, and on the other they hated him. But whether they loved or hated Paul, he was a success because he was doing what God had called him to do. You and I can do the same thing today. Both of us should.

FOLLOW THROUGH

DAY 268

Acts 15:36 And some days after Paul said unto Barnabas, Let us go again and visit our brethren in every city where we have preached the word of the Lord, and see how they do.

I remember my first day with my first child. There were cards, congratulations, and celebrations; but when Sena and I took our baby home from the hospital, everyone else went to their homes too. At three o'clock in the morning, no one was there to pat me on the back or change a diaper! I realized right then and there that this parenting thing would take *follow-through*.

Anything worth a stand is worth a follow-through. Living for the Lord will not always be easy, but that is why it takes dedication and stubbornness. The difference between what the Lord has done and what He would do, could do, and wants to do is largely a matter of follow-through. Paul and Barnabas went to visit the churches they helped to *"see how they do."* They knew the important work they had started necessitated a deliberate follow-through.

Has the Lord been working in your life? Have you taken steps in following Him that have demanded taking a stand? Take heart and be encouraged. If something is important enough to warrant a stand, then it is important enough to follow through! I know a pastor who was saved as a boy at camp thirty-five years ago and is now himself bringing kids to camp. He is following through with teenagers today because his pastor and church followed through with him three decades ago! That is worth the effort!

I am thankful for birth, but **birth is just the beginning.** Just like a newborn baby requires attention and follow-through, the good work that God has started in your life or in the lives of others has only just begun.

DAY 269

START AT HOME

Acts 16:34 And when he had brought them into his house, he set meat before them, and rejoiced, believing in God with all his house.

Acts 16 is the story of two preachers in a prison. Paul and Silas were placed in a prison; and God, as He had done with others before, opened the prison doors and freed them in a remarkable way. The jailer was scared out of his wits for good reason, and he came in before Paul and Silas and said, "Sirs, what must I do to be saved?" They answered, *"Believe on the Lord Jesus Christ, and thou shalt be saved, and thy house."* The question was "How can I be saved?" and their response to this man was, "Here's how you and your family can be saved."

That little part of the answer, "and thy house," is carried on through the next few verses. Verse 32 says, *"And they spake unto him the word of the Lord, and to all that were in his house."* Verses 33 and 34 say, *"And he…was baptized, he and all his, straightway. And when he had brought them into his house, he…rejoiced, believing in God with all his house."*

Paul wasn't going to settle for one man being saved when he could get the entire house. He wanted to win the world. You also see a glimpse of Paul's working assumption and God's bias regarding your home: **if you want to win the world to Christ, start by winning the people in your house.** Someone says, "Well, Brother Wil, the hardest people to win are the people in your own house." I understand that, but it should not be the case that you can't win people simply because they know you well. You should want the people who know you the best to respect you the most.

If I do right by the people in my house, my home will not be in competition with my ministry to people outside my home. Doing right at home can only strengthen my ministry to people outside of it. The question is not, "Am I going to be a good father or a great preacher?" or "Am I going to win the people in my house or the people on my block?" No. One is the foundation for the other.

All of us have problems in our homes, and all of us have things we wish we could do over again. But God's mandate in Acts 1:8 is to be witnesses where we are, first of all. Jesus said, *"... And ye shall be witnesses unto me both in Jerusalem, and in all Judaea, and in Samaria, and unto the uttermost part of the earth."* No one is closer to where you are than the people that share the inside of your house. Win the world, but begin by winning the people that are closest to you.

OFTEN IMITATED, ALWAYS DUPLICATED — DAY 270

Acts 20:18 And when they were come to him, he said unto them, Ye know, from the first day that I came into Asia, after what manner I have been with you at all seasons.

Did you know that you are the only person you know who has never seen you? You have never seen the way you look and walk in the same way that others see you. For instance, haven't we all heard ourselves on an audio recording and thought, "Do I really sound like *that*?" The easiest way to see yourself is to see the people who follow you.

You can only produce in your followers (kids, peers, neighbors) what you are producing in your own life. **You don't reproduce what you *want*; you reproduce what you *are*.** That is why Paul's statement in Acts 20:18 is so amazing! *"Ye know, from the first day ... after what manner I have been with you at all seasons."*

Perhaps no one knows you better day in and day out, *"at all seasons,"* than your family. You may be known for many things outside your home, but nothing will serve you longer and better than what you are and what you do inside your home.

Paul served with humility (verse 19), generosity (verse 20), and faith (verse 22). Do those traits characterize your life, your home, and your work? The heritage you leave will be visible in those who follow you. **You will reproduce tomorrow what you are producing today.** May God help us to be able to say with Paul, *"Ye know … what manner I have been with you at all seasons."*

DAY 271 — THE FAMILY OF GOD

Acts 21:17 And when we were come to Jerusalem, the brethren received us gladly.

I replied on Twitter to a missionary in Papua New Guinea who had asked for prayer. I was amazed when literally minutes later, he tweeted a reply! To the best of my memory, I've never met this friend; but I had an encouraging conversation with someone halfway around the world from the comfort of my Tennessee home! After our simple exchange, I was reminded that both of us need each other. He needs me to pray for his work in a foreign country, and I need him to pray for the ministry here at the Ranch. All of this was possible because we are both part of the family of God!

In a hostile world, it is wonderful to be in a family where God Almighty is the Father. There is fellowship along life's way, even from people you do not know. The same was true for the Apostle Paul in the book of Acts. As he journeyed to make and encourage disciples, *he* was encouraged!

Do not think that you can go through life alone. You need the encouragement of other Christians. We live in a hostile world that is opposed to everything good and godly. However, there is a God in heaven who is your Father, and He has many children that both give and need encouragement. **The best way to get the encouragement you are looking for is to *be* the encouragement you are looking for to someone else!**

Somewhere in Papua New Guinea, there is an American missionary who was encouraged by my prayers. The next time I tweet about some need I have, I am going to guess there is a missionary who will surely pray for me. That is how it is supposed to work in the family of God!

"But the end of all things is at hand: be ye therefore sober, and watch unto prayer. And above all things have fervent charity among yourselves ... use hospitality one to another without grudging." (I Peter 4:7-9)

◇◇◇

CITIZENSHIP IN HEAVEN

DAY 272

Acts 22:6 And it came to pass, that, as I made my journey, and was come nigh unto Damascus about noon, suddenly there shone from heaven a great light round about me.

I love to travel, and I love watching people in the airport. Any big airport has people from all corners of the earth going to every place imaginable. Even though I love the excitement of travel and the variety of people, I am thankful to be an American citizen. Citizenship frames so much of how we see ourselves and the way we view the world. The entirety of Paul's story in Acts 22 is about citizenship.

For instance, in verse 3, Paul says, *"I am verily a man which am a Jew, born in Tarsus, a city in Cilicia."* He goes on to talk about his Jewish credentials. Paul was part of God's special, precious, chosen people, the Jews, but that was not the primary citizenship that framed who Paul was and how he thought. In fact, many Jewish citizens, as noble as their heritage was, wanted to kill Paul because they believed he had betrayed that citizenship by trusting in Jesus as Christ. These people had rejected the Christ that God had sent.

Later on, the Romans were going to scourge Paul, and Paul claimed his Roman citizenship. Verse 27 says, *"Then the chief captain came, and said unto him, Tell me, art thou a Roman?*

He said, Yea." Paul's citizenship was very important because it protected him from an unjust scourging by these Romans.

Citizenships on earth tend to frame our prejudices and our expectations of our rights. We can be thankful if the rights we enjoy are protected by the citizenship we possess. Paul definitely made use of that. But Paul's life was not governed by his citizenship as a Jew or as a Roman; it was governed by a citizenship in heaven.

His testimony is found in verse 6 where he says, *"And it came to pass, that as I made my journey, and was come nigh unto Damascus about noon, suddenly there was shown from heaven a great light round about me."* He came to grips with Jesus, God's Messiah, who was sent to the whole world. The Lord Jesus told Paul that He had a plan for him to be a witness of what he had seen and heard.

A citizenship on earth frames our prejudices and rights, but **a citizenship in heaven frames our love of others and a surrender of self.** What citizenship most influences you and fosters, animates, and energizes your actions today?

◇◇◇

DAY 273
YOUR RESPONSE-ABILITY

Acts 28:6 Howbeit they looked when he should have swollen, or fallen down dead suddenly: but after they had looked a great while, and saw no harm come to him, they changed their minds, and said that he was a god.

Sometimes people love you; sometimes people hate you. Sometimes they can love you and hate you all in the same day! We can worry too much about what other people think about us, and that can cause us to lose our focus on God.

If you focus on following God, God can focus on the kind of influence you have on others. God can take care of someone's response to me if I rightly respond to Him. It is human nature to spend more time thinking about how others respond to me than how I respond to God.

Imagine being the Apostle Paul for one day in the book of Acts. You probably wouldn't have picked the day described in Acts 28! Shipwrecked and on an island with barbarians, Paul was bitten by a venomous snake. The barbarians were convinced he was a murderer but then changed their minds after he didn't die and called him a god. Talk about a roller coaster of emotion!

At the end of the day, your response to God is on your head—not anyone else's head. You can't blame your response on your parents, your environment, or your circumstances in the day. **Your response is your responsibility.** Don't get so distracted by who is following you that you miss out on following God like you should.

◇◇◇

I AM NOT ASHAMED

DAY 274

Romans 1:16 For I am not ashamed of the gospel of Christ: for it is the power of God unto salvation to every one that believeth; to the Jew first, and also to the Greek.

We are living in a day when American "exceptionalism" is being questioned. The idea that what we have as a country is uniquely wonderful and God-given has become a little embarrassing. We are now ashamed of what made us strong!

I can remember when Ronald Reagan was elected President. Morale was low, and in many cases people were ashamed to be Americans. In 1980, an advertising campaign called "Morning in America" ran on television. The ads were simple, morning scenes of everyday American life, and the idea of each ad was optimism, vision, and hope for America in the days ahead. The point was to return, not to arrogance, but to appreciation of a uniquely great country.

The mentality of being ashamed of our power is shared, not only by Americans, but also by Christians. We know that Jesus Christ is *"the way, the truth, and the life,"* but we are bombarded every day with tolerance for everything except this fundamental Bible truth. We can actually become a little ashamed of the power that we have! Paul said, *"I am not ashamed of the gospel of Christ: for it is the power of God unto salvation...."*

The strength we possess leaves no room for shame. The truth is, there is no gospel apart from Christ! The gospel is the power *"of God."* It is God's power, not ours, that is powerful enough to convince and save *"every one that believeth"*! It is sufficient to save anyone, from a drunk off the street to a little child, that will come to God through Jesus Christ.

Remember next time you are sharing the good news of salvation that the gospel is *"the power of God unto salvation to every one that believeth."* There is no room for shame or embarrassment in the gospel! I am proud to be an American, and I am proud to be a Christian. Never forget that the power you possess ***as a Christian*** is the power of Almighty God!

◇◇◇

DAY 275
BE WHO HE MADE YOU TO BE

Romans 1:21 Because that, when they knew God, they glorified him not as God, neither were thankful; but became vain in their imaginations, and their foolish heart was darkened.

Many people in this world are trying to find themselves while they have lost God. They are looking for the wrong person. **You cannot be who God made you to be until you acknowledge God for who He is.** That is where you must begin.

You can't be thankful for who you are if you resent who God is and who He made you to be. Either you will recreate who you are, or you will attempt to recreate the Creator. You cannot improve upon who God made you to be and what He made you to do!

Life is a grind when you try to be something or someone that God didn't make you to be, but **life is a privilege and joy when you live in submission to the Creator.** You'll never live higher than the god you have created. Any life lived outside of God's design is woefully inadequate, empty, and even wicked (read verses 24-32). When you live in light of who the Creator has made you to be, you have the resources and power of a limitless God!

◇◇◇

JUST AND THE JUSTIFIER

DAY 276

Romans 3:26 To declare, I say, at this time his righteousness: that he might be just, and the justifier of him which believeth in Jesus.

Have you ever done something that was wrong, but you tried to pass it off as right? Maybe you tried to convince yourself or someone else that what you did was OK. It's amazing the stories we tell ourselves and other people when we are trying to justify our actions!

Is justifying something or someone always a bad thing? What does it mean when God justifies a sinner? When you justify your wrong actions, you excuse your sin; when God justifies you, He deals with your sin. **Justifying yourself means you sweep your wrong "under the rug"; being justified by God means He puts your sin under the blood of Jesus Christ.** There is a cosmic difference between your act of justifying and God's!

The more you justify yourself, the less just you are. In contrast, **God is both just and the Justifier!** All mankind hinges on the truth that God is just and that He can justify. He is the only One who can be both at the same time!

God is just because He condemns sin, and He is just because He saves from sin. Any man, woman, boy, or girl who will be justified must come to the Justifier. If you are justified by faith in Jesus Christ, how could you boast or brag about any good that you have done? (verse 27)

In life, you can be just or you can try to justify your actions, but you can't be both just and a justifier. God *is* just and *the* Justifier, and we would do well to remember that in matters great and small.

DAY 277
YOUR CHOICE ABOUT PROBLEMS

Romans 8:31 What shall we then say to these things? If God be for us, who can be against us?

Do you anticipate any problems today? I am not saying that you would wish for them, but how could anyone expect a day without having any problems? Problems are part of life, and most problems are packaged as people. You may work with cars, paper, animals, or numbers, but none of these things give us as much anxiety as people problems do. People are by definition problems because all of us are by nature sinners; hence, there is conflict. So, you have a choice today. **Either you will wish for the absence of problems, or you will live in the presence of God,** acknowledging Him.

In Romans 8, God speaks of the sacrifice of His Son and the ministry of His Spirit. The Bible even speaks of the Holy Spirit helping us with things as basic as prayer. We can't even ask for help without help. Thankfully, God provides that help. God has given us His Son and His Spirit. So you will either wish for the absence of problems, which is wishing against reality, or you will acknowledge the presence of God.

Romans 8:31 says, *"What shall we then say to these things? If God be for us, who can be against us?"* Now if you give yourself ten seconds to think about that question, you probably can think of many answers. Who can be against us? Well, many people can be against us.

The truth is that no matter who may be against us today, if we are with God, then He is with us. If He is with us, then we are more than conquerors. We are not conquerors because we have no problems. You cannot be a conqueror unless you have a problem to conquer. But we are conquerors because God is with us. We are not conquerors because we are great but because God is, and He has provided everything we need to do as He wishes.

JESUS IS THE ISSUE

DAY 278

Romans 10:3 For they being ignorant of God's righteousness, and going about to establish their own righteousness, have not submitted themselves unto the righteousness of God.

In Romans 10, the Apostle Paul is speaking to people who were very religious but had rejected Christ. Their problem was what they had done with Jesus Christ, not what they had done with good works or with Paul. The issue was Jesus. In Romans 10:3 Paul says, *"For they being ignorant of God's righteousness, and going about to establish their own righteousness, have not submitted themselves unto the righteousness of God."* What were they rejecting? To what had they not submitted? It was not a what; it was a Who. The next verse says that the "Who" is Christ.

Verses 5 and 6 talk about the difference between the righteousness which is of the law and that which is of faith. The only righteousness that God is not duty-bound to reject as inferior is His own. The only way to have God's righteousness on your account is to trust God's Son as your Savior. When you talk to friends or counsel people, the problems may be complex, but the answer is not complex. The answer is simple and single. It is not a fact; it is a Person. It is not a 5-step plan. The answer is Jesus Christ—the way, the truth, and the life.

So, the issue is Jesus, not someone's pet sins, church, or creed. The issue is what he or she has done with Jesus Christ. So many people have trouble with sin because they have rejected Christ. In verse 21 Paul says, *"But to Israel he saith, All day long I have stretched forth my hands unto a disobedient and gainsaying people."* Against what did they rebel? It was not against what; it was against Who—God's Son.

Jesus is the issue. It matters not who you are. The problems may be complex today, but the truth is simple. It may be hard. It may be difficult, and it may go against the grain of human reason, but the truth is simple because the truth is a Person. The way is a Person. God's plan of salvation is in a Person. Remember, the issue is Jesus.

◇◇◇

DAY 279

THERE'S MORE!

Romans 11:33 O the depth of the riches both of the wisdom and knowledge of God! how unsearchable are his judgments, and his ways past finding out!

Some time ago, an oil rig in the Gulf of Mexico exploded, and the resulting millions of gallons of oil that spewed out captivated our nation and much of the world. Each passing day, men worked endlessly with the hope that the oil would stop, but it just kept coming! About that oil rig you could say, **"There is more where that came from."**

In the Pacific Ocean, there is a place called the Mariana Trench. It is nearly 36,000 feet deep—almost 7 miles! If you were in a submarine and on your way to the bottom, you could say, "There is more where that came from!"

If you are ever hiking out of the Grand Canyon on Bright Angel Trail, I have some good news and some bad news. The good news is, there is great ice cream at the top; the bad news is, looking back at the trail you've already hiked, there's more where that came from!

In the early summer, a thirteen-year-old boy summited Mount Everest. The summit is 29,028 feet elevation! Base camp at Everest is 14,000 feet. Most of us have never been to 14,000 feet, and that is only halfway up! Many a hiker has said on his way up Everest, "There is more where that came from."

The four examples above pale in comparison to God's supply. Romans 11 reminds us that there is more where that came from! God's wisdom and knowledge is never depleted. Have you ever needed wisdom that you didn't have? God's supply is deeper than your greatest need! What's more, His ways even defy our ability to comprehend. He truly is God!

You will never run dry if you are tapped into God. If you are tapped into yourself, you'll run dry before the day is done! God has all the wisdom, energy, and strength you lack. You would sooner drain the Mariana Trench with a straw than exhaust what God has available. I can't think of a better encouragement for today than verse 36: *"For of him, and through him, and to him, are all things: to whom be glory for ever. Amen."*

◇◇◇

BE HONEST WITH YOURSELF

DAY 280

Romans 11:36 For of him, and through him, and to him are all things: to whom be glory for ever. Amen.

What is the greatest accomplishment in your life? Whatever it may be, there is one of two ways you can feel about it. You can either feel great ("Hey, look at me"), or you can feel grateful ("Hey, thank the God Who gave me the ability to do what I did"). Now the latter statement may sound very pious, but it is also very true and practical. It is, in fact, the reality.

Romans 11:36 says, *"For of him, and through him, and to him are all things: to whom be glory for ever. Amen."* In other words, whatever comes from God belongs to God. If what I have done is of me, then I had better enjoy the "golf applause" I receive because that is all I am getting. If God is the One that has done it, then He is the One that deserves the glory. If God gave the gift, then God gets the glory.

In Romans 12, the Bible talks about the will of God and how God has gifted each of us. Maybe you don't know how God has gifted you, but He has. In Romans 12:3, Paul says, *"For I say, through the grace given unto me, to every man that is among you, not to think of himself more highly than he ought to think; but to think soberly, according as God hath dealt to every man the measure of faith."* God did not give the gift of ability or strength to enhance you. God gave it to enhance the work of Christ through you.

Now, this is the upshot of all this: **humility is little more than just being honest with yourself.** Sometimes, we can almost be proud of being humble, and we compare ourselves to other people. C.S Lewis said, "Pride takes no pleasure in having something, only in having more of it than the next man." In other words, if you take away comparison, pride has little to stand on.

Humility is looking at ourselves in relation to the Master that we serve, not in relation to fellow servants. God has given you your abilities and opportunities; and when you realize that God is the Giver, it changes the way you look at your work and your coworkers. I am to love them and thank God for gifting them so that together we can serve the Lord Jesus.

◇◇◇

DAY 281

ONE AND MANY

Romans 12:5 So we, being many, are one body in Christ, and every one members one of another.

When we trust Christ, the Bible tells us that God places us in the body *"as it hath pleased him."* (I Corinthians 12:18) I don't know where you are in the body of Christ, but I am thankful that we are not all doing the same thing! Notice the words *many* and *one* in Romans 12:4-5: "For *as we have **many** members in **one** body … we, being **many**, are **one** body…."*

Your physical body is one body, but it is made up of different parts like eyes, ears, toes, and lungs. They each have a specific purpose, and they are all important. As for which is more important, I always say that I want to keep everything I came with!

The important truth to keep in mind is that **God has placed you where it pleases Him.** He placed you where *He* wanted you, and you can't do better than that! You may not know exactly where He has placed you or what He has gifted you to do, but you can be faithful with what you *do* know right now. **It is wonderful that God never asks more or expects less of you than to be faithful.**

There is no room for pride or jealousy in the body of Christ; there is only room for gratitude to God for where He has placed you! He has placed you in a unique spot, in concert with others around you. Thank God for His care for you and for the fact that you are not in the body alone. As you serve the Lord, be thankful you can work with so many different "parts" and serve together as one body!

TO BE LIKEMINDED IS TO BE LIKE CHRIST

DAY 282

Romans 15:5 Now the God of patience and consolation grant you to be likeminded one toward another according to Christ Jesus.

Are you a patient person? If you need several minutes to answer that question, then you probably are. If the question irritates you, then probably you are not. Do you think God is patient? Romans 15:5 says, *"Now the God of patience and consolation grant you to be likeminded one toward another according to Christ Jesus."* How can a group of people who are so different in temperament, perspective, background, and so on work together and be likeminded? Must they all have the same personality, temperament, and perspective in order to work together? No. They must be *"likeminded one toward another according to Christ Jesus."*

Verses 6 and 7 go on to say, *"That ye may with one mind and one mouth glorify God, even the Father of our Lord Jesus Christ. Wherefore receive ye one another, as Christ also received us to the glory of God."* The only way for a group of believers to be likeminded is to be like Christ.

Philippians 2:5 says, *"Let this mind be in you, which was also in Christ Jesus."* It goes on to describe how God the Son was both humble and obedient to the Father even to the point of death. If God the Son exhibited humility and obedience, who am I to be a rebel and to be arrogant?

Romans 15:13 says, *"Now the God of hope fill you with all joy and peace in believing, that ye may abound in hope, through the power of the Holy Ghost."* Notice what the Bible is teaching us here. I am not the one who has patience; God is. I am not the one who is naturally filled with hope, but God gives hope. I am not the one who is naturally full of comfort, but God gives comfort. These are the virtues of the Father.

We also read of the example of the Son. The example we look to when we wish to be like God the Father is God the Son, because in Him dwells all the fullness of the Godhead bodily. So, the God of patience, consolation, and hope is exemplified in the life and work of the Lord Jesus Christ, God the Son.

The power is of the Holy Spirit of God. The last part of verse 13 says, *"through the power of the Holy Ghost."* So we have the virtues of the Father, the example of the Son, and the power of the Holy Spirit. The way to be likeminded today is to be like Christ through the Spirit's power.

GOD-GIVEN STEWARDSHIP

DAY 283

I Corinthians 4:7 For who maketh thee to differ from another? and what hast thou that thou didst not receive? now if thou didst receive it, why dost thou glory, as if thou hadst not received it?

Who do you know that is exactly like you? That's what I thought. Nobody is exactly like you! Who made you to be different from everybody else? God did. God is a God of diversity and variety. Being different (in this context) is not a bad thing; it is a good thing because God made us that way!

God has placed us in the body *"as it hath pleased him."* (I Corinthians 12:18) Our place in the body and the corresponding gifts have been given as a stewardship from God. Regarding this stewardship, God says in I Corinthians 4:2, *"Moreover it is required in stewards, that a man be found faithful."* **His standard is faithfulness—nothing more and nothing less.** So don't try to live up to a benchmark that He has not given to you! Often we make someone else's stewardship the standard. That's silly! God has given *you* a stewardship, and He requires faithfulness in that stewardship. **It would be wise to find out what your particular stewardship entails and then "go for broke"!**

If I think I am great because of my own efforts, that is *"wood, hay, stubble,"* and everyone will know it soon enough. (I Corinthians 3:12-13) However, any gifts I possess are a God-given stewardship; they are not mine to begin with! Since this stewardship is God's, I should give Him the glory and credit for what I do.

God made each of us unique, and He has given us everything we need to effectively serve Him. God has given people to preach, to serve, to teach, and to encourage and comfort. Since we are each given a unique stewardship, we can rely on the fact that we can do what we should do. That's true of everyone here on the Bill Rice Ranch, and that's true of *you*!

DAY 284

THE WORST MISTAKE TO MAKE

I Corinthians 10:6,11 Now these things were our examples, to the intent we should not lust after evil things, as they also lusted.... Now all these things happened unto them for ensamples: and they are written for our admonition, upon whom the ends of the world are come.

What is the worst mistake you have ever made? I may not know much about you, but I do know the worst mistake you can make is the one you make the second time. Have you ever said, "Oh no, not again!" or, "I can't believe I did that again"? The following are three helpful ways to learn from mistakes.

First, the Apostle Paul encourages us in chapter 10 to **learn from life itself.** God does not want us to be ignorant (verse 1)—ignorance is not bliss! You don't have to be intelligent to learn from life. Use each trial, success, and failure as a learning experience. Your life is but a vapor (James 4:14), so learn what you can from each day and moment.

Secondly, we ought to **learn from our own mistakes and the mistakes of others.** You have more history from which to learn than any generation before. How are you putting that experience to use in your life? We have the privilege of learning from thousands of years of human mistakes. How tragic to repeat others' mistakes when you could have learned from them!

Thirdly, we should **learn from the Bible**. That sounds obvious, but notice again verse 11: *"And they are written for our admonition...."* From the narratives to every psalm to every epistle, the Bible is a help in every facet of life. God's Word was not written in a vacuum, only to be used and heeded by the first recipients. You can learn so much from the Bible! Every psalm has a story, and the New Testament epistles are great to learn from because they were often written to correct what was wrong.

This is a week to learn from your mistakes—today is a great day to start. Analyzing and reflecting on your work after a period of time can be beneficial. Don't squander an opportunity to learn from your mistakes. When you think about mistakes, don't be deceived—the wrong attitude will always lead you to the wrong place (verse 12); don't be dense—staying ignorant is not smart; and don't be defensive—be honest about yourself and others.

Take time this morning to think about your week, month, year, etc., and learn from your mistakes and the mistakes of others. **The worst mistake is the one made twice.**

FOOD FOR THOUGHT

DAY 285

I Corinthians 10:31 Whether therefore ye eat, or drink, or whatsoever ye do, do all to the glory of God.

What did you eat this morning? There is probably a wide range of answers, but let me ask you a second question: **why did you eat this morning?** Was it for fun? For fellowship? For fuel?

The book of I Corinthians is a book full of food! Notice our verse this morning which says, *"Whether therefore ye eat, or drink…."* In I Corinthians 5:8, Paul speaks of the *"leaven of malice and wickedness"* and the *"unleavened bread of sincerity and truth."*

In chapter 8, Paul addresses questions concerning meat offered to idols. Meat at someone's home, at the market for purchase, and at the temple with idols all had a definite Bible answer: *"do all to the glory of God."* In chapter 11, he deals with those who were partaking of the Lord's Supper irreverently (verse 20). These folks were thinking about everything except glorifying God!

But what is the reason Paul mentions food specifically so many times? For what reason should we be concerned with food this morning? *"Whether therefore ye eat, or drink, or whatsoever ye do, do all to the glory of God."* **Honor God in everything—including food!**

I do not know why you will eat later today, whether it is for fun, for fellowship with friends, or for fuel; but I do know that you ought to honor God in it. On occasion, we have banquets here at the Ranch. We have delicious food at these banquets. But the temptation is to think about the food or the logistics of a banquet and forget Who should be honored in it all! Honor God when you eat; honor Him when you set up and prepare for a meal; honor Him when you entertain guests. Do all to the glory of God!

Now, if you will excuse me, all this talk about food has made me hungry!

DAY 286
DIFFERENCE AND UNITY

I Corinthians 12:4-6 Now there are diversities of gifts, but the same Spirit. And there are differences of administrations, but the same Lord. And there are diversities of operations, but it is the same God which worketh all in all.

I find it interesting that God says in verse 1 of this chapter, *"Now concerning spiritual gifts, brethren, I would not have you ignorant."* Ignorance is not a good thing; it is not a virtue to be unaware. In verses 4-6 you see two themes, difference and unity: diversities—same Spirit; differences—same Lord; diversities—same God. Different people have different spiritual gifts, but you can't escape the word *same* in this passage!

Differences (in spiritual gifts) can either be complementary or contentious. God designed these differences; so don't live your life trying to be like someone else. Don't live your life without the power that comes with unity.

Often people can be misinformed about this deal of difference and unity. The way for a husband to be a good husband is not to be a good wife; he needs to be a good husband! The same is true for an evangelist—the way to be a help to a good pastor is to be a good evangelist. God made husbands and wives differently (evangelists and pastors too) in order for them to compliment one another.

Here are a few observations regarding spiritual gifts:

Gifted people are gifted where they are gifted. That is, everyone is gifted in a specific area (or areas). A person may be strong in one area and weak in another. Where a person is gifted is determined by God—the Father and Creator knows best.

When people are in the right spot, they can shine. The fable of the "Ugly Duckling" comes to mind here. Being in the right spot can make a world of difference when it comes to spiritual gifts! Spiritual gifts are all about serving the right Master in the right spot.

Something is not right if my serving is not profitable (verse 7). God is the One who designed the diversity that exists in spiritual gifts. Do give all you have to God and be on the same page. There really can be both difference and unity in God's family!

◇◇◇

THE GIFT THAT KEEPS ON GIVING

DAY 287

I Corinthians 12:7 But the manifestation of the Spirit is given to every man to profit withal.

Over the past several years, my son has received some exciting and wonderful gifts from his grandpa. He received a couple of pocket knives, a high-powered pellet gun, and some other dangerous items. These gifts are wonderful because they keep on giving! Whenever he is with Grandma and Grandpa, he can worry them with knives and guns!

The gifts that God gives us are much the same—they are gifts that keep on giving. The gifts God gave you should profit others, not just yourself. The point is not *what* gift you have; rather, it is what you are *doing* with what you have. In other words, **whatever gift God has given you is only as good as its profit to others.** It is true that *"a man's gift maketh room for him"* (Proverbs 18:16), but your gift was not primarily given just to make a way for you. The gifts God gave are not *to* you but *through* you.

You may have a different gift than I do, but both our gifts came from the same Source. My gift is not better than yours, nor is it a lesser gift than yours. The gifts came from *"the same God which worketh all in all."* (I Corinthians 12:6) **Nothing is better than a God-given gift!** If you are not sure what your gift is, start serving somewhere and you will find out soon enough. After all, God gave you the gift to *"profit withal"*!

◇◇◇

DAY 288 — TRUE LOVE

I Corinthians 13:4,7 Charity suffereth long, and is kind; charity envieth not; charity vaunteth not itself, is not puffed up ... Beareth all things, believeth all things, hopeth all things, endureth all things.

Corinthians 12 is about what God has given us (spiritual gifts); chapter 13 teaches us how we should give what we've been given. **Charity is what I give to others.** This well-known chapter about charity (love) is so refreshing and so thought provoking.

For example, verse 4 says, *"Charity suffereth long, and is kind...."* At the end of the day, it may be easy to say that you've been kind. But have you been kind *after* you have suffered long? That is a horse of a different color and much more difficult! The verse continues by saying that charity does not envy or puff itself up. That is enough to keep me busy for a while; and that is only one verse!

Haven't we all been to a restaurant where the employees do not "buy in"? Poor service, apathy, and laziness overshadow the food, no matter how good it tastes! **What you have is God's; what you do with what you have is your responsibility.** You need to "buy in" to God's type of love. How does your charity rate on God's scale?

Verse 5 gives a couple of good points about charity to consider, as well. *"Is not easily provoked"* and *"thinketh no evil"* apply to our typical conversations. Are you easily "set off" (easily provoked)? Have you ever stopped listening to what a person was saying because you already knew what you were going to do anyway (thinking evil)? My, how I Corinthians 13 would change our lives if we put it into practice!

Do you think God is ever impressed by what *He* has given us? That would be silly, wouldn't it? The gift was His to begin with, and He is the One who gave it to us. How foolish to think that God will be impressed by His gifts! **God is looking for Christians who give what He has given to them. This is true charity—I Corinthians 13 charity.**

True charity *"beareth all things, believeth all things, hopeth all things, endureth all things."* May God help each of us to give to others in the way we should today.

Everything I have that is worth anything comes from God. It matters not what I have; it only matters what I am doing with what I have. The only good your gift has is the good it is doing. So what are you doing with the gifts God has given *you*?

◇◇◇

HOW STRONG IS YOUR FAITH?

DAY 289

I Corinthians 15:2 By which also ye are saved, if ye keep in memory what I preached unto you, unless ye have believed in vain.

How strong is your faith this morning? How could you answer that question definitively? Would you say, "I've got a 5.6 faith on a scale of 1-10," or "I have a 9.0 faith on a scale of 1-10"? How do you answer a question regarding the strength of your faith?

I Corinthians 15 gives us a clue about guaging the strength of our faith. Paul was talking about the power of the gospel, that

Christ died for our sins, was buried, and rose again. In verse 2 he says, *"By which also ye are saved, if ye keep in memory what I preached unto you, unless ye have believed in vain."* That conditional word "unless" is a little unnerving. It is almost as if Paul says, "Oh, good news? Christ died for our sin, was buried, and rose again. You can be saved—unless you have believed in vain."

Is it possible to believe in Jesus in vain? Only theoretically—that is the honest answer. Verse 14 sums it up this way, *"And if Christ be not risen, then is our preaching vain, and your faith is also vain."* If Jesus is dead, faith is too. If Jesus is dead, faith is empty. If the tomb is full, our faith is useless.

Verse 17 says, *"And if Christ be not raised, your faith is vain; ye are yet in your sins."* Verse 19 takes it a step further, *"If in this life only we have hope in Christ, we are of all men most miserable."* Now I know sometimes we say that even if Jesus were dead and Christianity were a fraud, living the Christian life is still the best life there is. I can appreciate the sentiment, but that is simply not true. If Jesus is dead, we are miserable. Our faith is vain. Our preaching is empty.

In verse 20 Paul turns the conditional "if" into a definitive "is." He says, *"But now is Christ risen from the dead, and become the firstfruits of them that slept."* Sometimes we talk about "easy believism," and I understand the concern people who use that term may have. But either you are believing in Jesus, or you are not. Easy believism is an unfortunate choice of words because its counterpart would naturally be "hard-believism," which would be works. I can't save myself by my works. I am saved because of Jesus' work. When I depend upon Him completely, this faith is the power of God to save my soul.

It is a powerful faith that rests in a living Christ. I don't know your need; but I do know that if your faith is in Jesus, you have reason for hope, power, and victory because Jesus Christ lives. Faith in Christ is not empty; it is powerful because He is.

A PROBLEM OR AN OPPORTUNITY?

DAY 290

I Corinthians 16:9 For a great door and effectual is opened unto me, and there are many adversaries.

Sometimes we are so intent on waiting for our ship to come in that we don't see it when it does. We are hoping for a cruise liner, but our ship comes in as a dirty old barge full of boxes of problems. **The difference between a royal pain and a golden opportunity is often how you see it.**

In I Corinthians 16, Paul was talking about where he was going and what he was going to do. He assessed everything in verse 9 when he said, *"For a great door and effectual is opened unto me, and there are many adversaries."* This great and effectual door was opportunity. The adversaries were opposition.

Sometimes, problems are opportunities that find you. Problems and opportunities don't cancel out one another; they come together. Now, two opportunities can cancel out one another. If I have a given situation, I can only take one of a number of opportunities. I can't do everything. I must choose. But having a problem does not exclude having an opportunity also. Sometimes an opportunity comes to me as a problem, and almost always opportunities come with problems.

Let's say that someone is given a thousand acres. They say, "Wow! I have a thousand acres. How can I serve God with this? I think I will start a camp." I admire that. I thank God for a person that thinks about what they have been given in light of how they can serve the Lord, but don't be discouraged if God does not give you a thousand acres. That doesn't mean you don't have opportunity. The Ranch is not here because Bill and Cathy Rice were given a thousand acres. They were given a daughter who became deaf. They were given problems that seemed insurmountable.

Looking back, we can see now that the greatest tragedy of Bill and Cathy Rice's lives was also God's greatest opportunity and calling for them. That is why the Bill Rice Ranch is here today.

Don't be discouraged if your golden opportunity looks like a royal pain. **When we see problems, God sees potential.**

◇◇◇

DAY 291 — THE GOD OF ALL COMFORT

II Corinthians 1:3-4 Blessed be God, even the Father of our Lord Jesus Christ, the Father of mercies, and the God of all comfort; Who comforteth us in all our tribulation, that we may be able to comfort them which are in any trouble, by the comfort wherewith we ourselves are comforted of God.

Recently, I was reminded that the home so often frames and illustrates who God is to us. That makes the family incredibly important! **You probably think of God's care and correction in the framework of someone in your family.** II Corinthians 1:3-4 describes God as *"the Father of our Lord Jesus Christ, the Father of mercies, and the God of all comfort."* What descriptions of His character!

God is *"the God of all comfort,"* and we learn in II Corinthians 1:4 that **He primarily comforts through people.** He comforts us so that *"we may be able to comfort them which are in any trouble, by the comfort wherewith we ourselves are comforted of God."*

In II Corinthians 7, we find an interesting story about the Apostle Paul. He needed comfort after experiencing trials, both inward and outward. What happened? *"Nevertheless God, that comforteth those that are cast down, comforted us by the coming of Titus."* Was it God or Titus that comforted Paul? The answer is, yes! God is the God of ALL comfort—there is no comfort apart from Him. Equally true is the fact that Titus was the face of God's comfort!

Why are you going through tribulations? I cannot pretend to know all the reasons, but **one of the good things about adversity is the fact that the comfort you receive is the very same comfort God can bring to someone else's life through you.** God is a good God—He is a good Father. As frail as our human families are, they can remind us of the truth that our Father is the God of all comfort. Are you using the comfort you receive to comfort others? You just might receive comfort today that someone else will need tomorrow!

AIMING FOR THE ETERNAL

DAY 292

II Corinthians 4:18 While we look not at the things which are seen, but at the things which are not seen: for the things which are seen are temporal; but the things which are not seen are eternal.

What makes more sense—investing in something temporary or investing in something permanent? You can spend money remodeling your home, and you can expect a return on your investment. On the other hand, I've heard people talk about "investing" in an R.V. or fifth-wheel trailer. I can tell you from personal experience that a fifth-wheel is not an investment! It may serve your needs well, but a trailer never appreciates in value. Parts wear out, things break, and you never recoup the money you put into it!

We may not aspire to own a "home on wheels," but just think about all the time and money we "invest" in things like food, clothes, and personal items. On top of that, all these things go into maintaining our bodies which will not last! A pantry full of food will be empty in a heartbeat (ask any parent of a teen guy); clothes will wear out and go out of fashion; and toothpaste, soap, and deodorant are a daily need! **None of these things are bad, but none of them will last.**

What you and I have right now is like a tent; it is temporary. What we have awaiting us in heaven is like a building; it is eternal! Each day we ought to aim for what is eternal. **Most of us fret our lives away about the urgent instead of the important.** There is a difference between living in light of the important and living in light of the urgent.

II Corinthians 4:18 says, *"While we look not at the things which are seen, but at the things which are not seen: for the things which are seen are temporal; but the things which are not seen are eternal."* How do you look at things which are not seen? You **"aim for"** them. You have an opportunity every day either to aim for eternal matters or to focus solely on the here and now.

It is not wrong to have a job, a car, a house, or kids that play soccer, take tuba lessons, or any number of things. What is wrong is to be so caught up with keeping up with the neighbors or following the latest trends that you forsake what is eternally important. Investing in the eternal may require investing in temporary things, but that should never take the focus off of the ultimate goal.

What are you aiming for this morning? Is your life busy with temporary things or eternal things? To focus on eternal things requires vision and wisdom that can only come from God. May it be true of my life and your life today that we are aiming for eternal things.

◇◇◇

DAY 293
YOU ARE NOT ALONE

II Corinthians 6:1 We then, as workers together with him, beseech you also that ye receive not the grace of God in vain.

Increasingly, we live in a hostile and lonely world. Because of technology people have more access to others than they have ever had before. Yet, people seem more alone, more marginalized than ever before. You see subway cars, buses, and buildings full of people caught up in their own individual worlds with their eyes on their phones, their ears plugged with an iPod or some other device. It seems like we are increasingly lonely people. Well, if you are serving God, you are not alone. You need to remember that.

In II Corinthians 6:1 Paul says, *"We then as workers together with him, beseech you also that ye receive not the grace of God in vain."* So, we are workers together with the Lord. Paul says, *"I beseech you."* For whom does he beseech these people? If you look at II Corinthians 5:20 he says, *"Now then we are ambassadors for Christ, as though God did beseech you by us: we pray you in Christ's stead, be ye reconciled to God."* So God is working through us, and God is working with us. We are working for Him. II Corinthians 6:4 says, *"But in all things approving ourselves as the ministers of God."* So, we serve "for Christ," "with him," "of God," and "by the Holy Ghost." In every way, when we serve God, we are not alone.

The Bible tells us later in the chapter that we are not to be unequally yoked together with unbelievers. The counter to that is that we are yoked together with the Lord Jesus. Matthew 11:30 says *"For my yoke is easy, and my burden is light."* What a wonderful truth that **when you serve God you never serve alone!**

◇◇◇

YOU ARE CONTAGIOUS!

DAY 294

II Corinthians 7:13 *Therefore we were comforted in your comfort: yea, and exceedingly the more joyed we for the joy of Titus, because his spirit was refreshed by you all.*

Have you ever had a cold? The only thing worse than *having* a cold is getting a cold *from* someone else! Like a cold, **your attitudes and actions are contagious.** What are you spreading? Even a family can have a dominant attitude as a group! What is your family's attitude?

Our attitudes will spread like a contagious cold. If you notice that you do not like the attitudes you find around you, stop looking *around* you and look *within* you! **If you don't like the attitude floating around, change it!**

It really is true that **attitudes are caught, not taught.** That can either be positive or negative, depending on what you are spreading. Just look at Paul and the believers in Corinth. The joy and comfort they received were products of the joy and comfort they were catching and spreading to each other!

If people in your workplace, your home, or your school catch what you have, what will they "come down with" today? Will they catch comfort and joy, or will they catch a wrong attitude? We ought to be conduits spreading what is right, good, and godly to others. The next time your family passes around a cold, think about what else you are spreading to those around you!

◇◇◇

DAY 295
THE FOOLISHNESS OF COMPARING

II Corinthians 10:12 For we dare not make ourselves of the number, or compare ourselves with some that commend themselves; but they measuring themselves by themselves, and comparing themselves among themselves, are not wise.

It is amazing how often we look at other people to either justify or motivate what we are doing. As a kid, did you ever say, "But, Dad, *they* get to do it"? Even though I tried that excuse, my dad never seemed motivated by what the preacher's kids (or anyone else's kids) were doing!

II Corinthians 10:12 gives great wisdom to live by: **don't ever compare yourself with those who commend themselves.** Someone's great appearance is not always an accurate reflection of reality. It is so easy to see people and think that they have it easier or better than they actually do. You never know all of the dynamics that are going on in someone's heart or home. To drive yourself crazy trying to keep up with someone else is not wise. That is exactly what the Bible says!

It is wise to heed the godly advice and examples of those who have done well (in raising their kids, for instance). However, always remember that **comparing yourself to someone else is never helpful and always foolish!** Our opportunities, our ministries, and our families are different from one another. Don't forsake what you've been given in order to chase after what someone else has been given.

Verses 16 and 17 are a great reminder to us this morning: *"…and not to boast in another man's line of things made ready to our hand. But he that glorieth, let him glory in the Lord."* God has gifted us all and given us all different opportunities. When all is said and done, **the only mark that matters is the one God sets for *you*.**

◇◇◇

ACKNOWLEDGE GOD IN WEAKNESS

DAY 296

II Corinthians 12:10 Therefore I take pleasure in infirmities, in reproaches, in necessities, in persecutions, in distresses for Christ's sake: for when I am weak, then am I strong.

Would you rather brag on your strength or acknowledge your weakness? That is pretty easy to answer, isn't it? All of us would rather talk about how strong we are than acknowledge how weak we are. It's not wrong to have strength, but it is wrong to have weakness and not acknowledge it. All of us are weak. We are just not always smart enough to know it.

Paul refers to his strength, his weakness, and his accomplishments in II Corinthians 11:30 where he says, *"If I must needs glory, I will glory of the things which concern mine infirmities."* In the next chapter he talks about some things for which he would have great cause to boast; but instead he says, *"Yet of myself I will not glory, but in mine infirmities."* In II Corinthians 12:6 he goes on to say, *"For though I would desire to glory, I shall not be a fool."* All of us have the tendency to brag, and we are fools when we do. **God empowers those who acknowledge Him in their weakness.**

God allowed Paul to have a difficulty that Paul called a thorn in the flesh. Paul indicates that God allowed this thorn precisely to keep his focus and dependence on the Lord. He asked God to take away this difficulty, but the Lord responded, *"My grace is sufficient for thee, for my strength is made perfect in weakness."* Then Paul says, *"Most gladly therefore will I rather glory in my infirmities, that the power of Christ may rest upon me…for when I am weak, then am I strong."*

Only the weak need help. Only the weak ask for help. Only the weak get help. Here is a little secret—all of us are weak. We are just not always smart enough to know it. You will live in your weakness as long as you settle for your own strength. We acknowledge people for their strength, but God empowers those who acknowledge Him because they are honest about their weakness and because they are dependent upon God's strength.

◇◇◇

DAY 297 — WHAT'S YOUR SOURCE?

Galatians 1:11-12 But I certify you, brethren, that the gospel which was preached of me is not after man. For I neither received it of man, neither was I taught it, but by the revelation of Jesus Christ.

About seventy years ago people seemed to have more regard for advertising and news than we do today. Maybe we were more simple in our approach to life or in our acceptance of advertising and news in days gone by. People are much more skeptical today, as well they should be. Many highly regarded news outlets have had their share of scandal, of plagiarism, and falsehood. Advertising is taken with a grain of salt today. The bottom line is that **the news you give and the news you receive is only as good as the source that you cite.**

In Galatians 1, Paul was refuting people who had another gospel and had put a twist on God's plan. He says in verse 11, "But I *certify you, brethren, that the gospel which was preached of me was not after man. For I neither received it of man, neither was I taught it, but by the revelation of Jesus Christ."* The gospel is not just a plan or schematic. The gospel is a Person. That is why Paul says in verse 6, "I marvel that ye are so soon removed from him that called you into the grace of Christ unto another gospel." These people had not just turned from God's plan; they had turned from God. They had not just rejected the gospel; they had rejected God's Son. The gospel is a Person.

Remember to keep the gospel simple. Must a person receive, believe, repent, and accept in order to be saved? Well, yes, but those are not numerous decisions. There is only one decision. There is only one sin that damns, and that sin is rejecting Christ. There is only one virtue that saves, and that virtue is receiving Christ. There is only one decision that stands between a person and eternal life, and that decision is what they do with God's Son. To reject the gospel is not merely to reject a doctrine, dogma, or plan; it is to reject a Person. Keep the gospel simple.

Also, keep the gospel about Christ. He is the single issue of the ages. He is the single issue standing between us and the eternity beyond. We can rest in Him and proclaim Him in all His simplicity and sufficiency.

◇◇◇

ONE GOSPEL TO EVERY SINNER

DAY 298

Galatians 2:8 For he that wrought effectually in Peter to the apostleship of the circumcision, the same was mighty in me toward the Gentiles.

Recently on the Ranch, we had three different services going on at the same time. We had services for teen, junior, and deaf campers. To the deaf campers we preached in American Sign Language; to the teenagers we preached in English; and to the juniors we preached in Junior! All joking aside, those are three different groups, and within those groups there are people with a wide variety of different backgrounds.

How encouraging to remember that there is but one gospel to every sinner, whether the person is Deaf or hearing, Jew or Gentile, young or old, or from Paul's day or ours. **There is only one gospel to every sinner.**

Paul said in Galatians 2:8, *"(For he that wrought effectually in Peter to the apostleship of the circumcision, the same was mighty in me toward the Gentiles.)"* Peter was sent primarily to take the good news to the Jewish people, and Paul was sent primarily to take the good news to the Gentiles. The same God that committed the gospel to Peter committed the gospel to Paul. In both cases God gave the gospel; God owns the gospel; God crafted the gospel; and it is our job to give it.

Presentation, on the other hand, requires flexibility. When it comes to juniors, I am going to preach to them in a way that they can understand the gospel; but I am not going to change the actual content of the gospel. When I preach to deaf people, I would use American Sign Language. If I knew the language, I could preach to Germans, Italians, or any one of a wide variety of peoples. There is even a wide variety of people among those who speak English. But we need to remember that there is only one gospel for every sinner.

God will give grace and the help we need. Tailor the presentation, but don't budge on the truth. The same God that empowered the greatest preacher you have ever heard is the same God that will work through you mightily if you will take what He has said, give it as it is, and give it to the people to whom God has given you access today.

LEAD TO VICTORY

DAY 299

Galatians 4:11-12 I am afraid of you, lest I have bestowed upon you labour in vain. Brethren, I beseech you, be as I am; for I am as ye are: ye have not injured me at all.

Perhaps the two times surrounding camp that contrast most with each other are the last night of camp and the ride home on the church bus. The last night of camp is a high point. God has worked in hearts; people are excited; young people are wide awake. And then you've got the bus ride back home. It can often be a long, boring trip. During this time, people begin to worry that the decisions at camp will not last when the normal routines of life kick in. I think the adults who bring young people to camp wonder, "Will decisions from camp last?" That is a valid question, and I believe it's a question that Paul would probably have wondered, as well, if he were here today.

In Galatians 4, Paul speaks of people in whom he had invested much. In verse 11 he says, *"I am afraid of you, lest I have bestowed upon you labour in vain."* Most of us do not mind working or expending energy, but none of us want to waste those resources of life. How do such workings of God last?

The way to see others go on to victory is to lead them there yourself. In verse 12, Paul addresses his own concern, *"Brethren, I beseech you, be as I am..."* Paul was a Jewish man, but he put aside everything he had previously hoped in and depended completely on the Messiah, the Savior of the entire world, both Jew and Gentile. **The way to help others be where they need to be is to lead them there by your example.**

We lead people through the work of the Son and by the power of the Spirit. In verse 4 Paul says, *"But when the fulness of the time was come, God sent forth his Son."* Jesus is better than a law. Jesus is better than a custom. Jesus is better than tradition. Jesus Christ has the power to save sinners and to transform their lives. We lead through the work of the Son, not merely by the labor we invest in others.

We lead by the power of the Spirit. In verse 6 he says, *"And because ye are sons, God hath sent forth the Spirit of his Son into yours hearts, crying, Abba Father."* Jesus is our Savior; God Almighty is our Father; and the Spirit of God is our Comforter. The Spirit enables us to lead by example.

Remember, the best way to lead others to victory is to live in victory yourself.

DAY 300

WHO'S DRIVING?

Galatians 5:16 This I say then, Walk in the Spirit, and ye shall not fulfill the lust of the flesh.

On a recent trip to the Carolinas, Sena and I were meeting up with family members at a local park. As we pulled in behind their minivan and followed them through the parking area, I noticed that the van was weaving back and forth. I could see the driver's face reflecting in the mirror, and all I could see was a big, cheesy grin! We were barely moving along, but that minivan was obviously under the influence of something. Well, as it turned out, that influence was my five-year-old niece!

Just as I could tell that van was under the influence of a child by the way it was tracking, **you can tell who is steering your life by the way your life is tracking.** Whose hands are "steering" your life? Who is calling the shots? Galatians 5 describes the epic battle between the flesh and the Spirit. If God's Spirit is driving my life, then I will do what I should!

My flesh can only do selfish, fleshly things. Only God's Spirit can do what He would do. I don't naturally, in my flesh, love people as I ought. I don't naturally, in my flesh, bear the fruit of the Spirit (Galatians 5:22-23). Only God's Spirit with His hands on the wheel of my life can do that!

Does love describe your life? Does joy describe your life? How about peace? Longsuffering? Gentleness? Goodness? Faith? Meekness? Temperance? If not, **take a look at whose hands are on the steering wheel!**

◇◇◇

WEARY IN WELL DOING

DAY 301

Galatians 6:9 And let us not be weary in well doing: for in due season we shall reap, if we faint not.

Have you ever gotten a belated birthday present? You know, of all the times I've received a gift for my birthday, I have never once turned down the gift! Have you ever grown tired of people saying, "Thank you"? Do you ever feel frustrated on pay day because you got a paycheck last week too?

Isn't it amazing how we never tire of receiving things, great or small? We never grow weary of a surprise gift, a grateful "thank you," or the wages we have earned. But have you ever been weary? If you are honest with yourself, **you probably were weary when you were giving!**

But remember, when you see to it that you give, God will see to it that you receive. That is how it works. You will reap *"in due season,"* like a farmer who reaps a crop at harvest. You don't get corn the day after you plant, but if you plant corn, you *will* reap corn. The key is to not faint or grow weary!

You are probably like me in that you are never weary of receiving. I love "getting," but I can grow weary in giving very quickly! God sees the beginning from the end; and you will reap, if you will not be weary.

"For in due season we shall reap, if we faint not."

DAY 302

SEASONS

Galatians 6:9-10 And let us not be weary in well doing: for in due season we shall reap, if we faint not. As we have therefore opportunity, let us do good unto all men, especially unto them who are of the household of faith.

Weary. **Well-doing**. Those two words just seem to go together, don't they? The Bible encourages us to *"faint not"* and to *"not be weary"* in doing well because there is a promise—*"for in due season we shall reap."* But the promise of reaping what is right is conditional upon not fainting!

Our nature never tires of serving self and serving the wrong thing, but sowing to the flesh will mean reaping corruption (verse 8). To reap good things, we must keep up the well-doing and not be weary or faint. A season of reaping is coming (verse 9).

In light of sowing and reaping, verse 10 mentions "opportunity," which is actually the same word as "season" in verse 9. There will be different opportunities—"seasons"—in which to do good. And the opportunities I have today and this month are probably different than the opportunities that were present yesterday and last month.

What are we to do? *"Do good."* **When** are we to do good? *"As we have therefore opportunity* (season).*"* **To whom** are we to do good? *"All men, especially unto them who are of the household of faith."* Verse 10 covers it all!

How do you start doing good to *"all men"* and to the *"the household of faith"*? Well, why not start with those in your own house?! How about the folks in your church? They certainly qualify! These are at least two places to start; anyone can begin there.

Just as a farmer knows what season it is—plowing time, sowing time, growing time, or harvest time—**we ought to be tuned in to the "seasons" we have now.** What you are going to do, do now. Make your work and time count. Don't worry about what other people may be doing or what other seasons may bring; consider what "season" it is today and get to work.

"And let us not be weary in well doing: for in due season we shall reap, if we faint not."

WHAT BRAND ARE YOU? DAY 303

Galatians 6:17 From henceforth let no man trouble me: for I bear in my body the marks of the Lord Jesus.

Here on the Bill Rice Ranch, we have over forty head of horses. Back in the early days of the Ranch, we would brand each horse so it was obvious who owned the horse. Isn't it fascinating how consumed we are with brands? Brands of clothes, brands of cars, and probably scores of other brands—brands mean more to us than we would care to admit!

What brand are you looking for this morning? **What mark makes a person an accredited servant of God?** Paul said, *"For I bear in my body the marks* [literally, "the brand"] *of the Lord Jesus."* We can often deal with people in the same way that we deal with brands. We make judgments based upon the brands we identify with, like Ford cars, certain name brand shirts, or sports equipment. Likewise, we often identify servants of God with *our* brands instead of His.

In any given church (or week at camp), a variety of colleges and backgrounds are represented. If you are not careful, you will look for the mark *you* would stamp on a person, not the mark that comes from serving Christ. There is more that could be said about right versus wrong and worldly versus godly; but the point is, we can put stock in the "brand" we like or identify with instead of the mark God has stamped on someone.

Paul's marks were both physical marks (scars) from shipwrecks, beatings, and imprisonment, as well as spiritual marks of his ministry. **The mark that brands a servant of God is the act of serving God!** A servant of God will not glory in himself and will

WIL RICE IV

be crucified to the world (verse 14). Regardless of what other brand (your college, your church, your ministry) you wear, the most important brand is the brand that God stamps! That will mean serving in humility and being dead to the world.

The mark we should be looking for in a servant of Christ is the attribute of *serving Christ!* Serving Christ means serving people and serving under the authorities He has placed in your life. Is your "brand" one that is glorying in Christ, crucified to the world, and representing the *"new creature"* you are? **What "brand" most marks you?** Don't look for the brand of your school, your interests, or your ministry; look for the one brand that matters most—service to Christ!

◇◇◇

DAY 304 — IN OVER YOUR HEAD?

Ephesians 1:22 And hath put all things under his feet, and gave him to be the head over all things to the church.

Have you ever been in over your head? Peter certainly knew what that was like. You may recall that he stepped out of a boat that was sitting on a sea full of water, took a few steps, and then was promptly in the water over his head. Of course, he was answering the call of the Lord Jesus Who had that water under His feet, Who had that nature under His control, Who had spoken the worlds into existence, and Who controls it even now.

Ephesians 1:22 says, *"And hath put all things under his feet, and gave him to be the head over all things to the church."* Just like Peter, I sometimes find myself in over my head. I have problems, challenges, and questions that overwhelm me; but nothing ever takes the Lord Jesus by surprise, and nothing ever exceeds the power of heaven. I am comforted to remember that the things that are over my head are still under Christ's feet. That is true for you, as well. **The things that are over your head are under Christ's feet.**

That truth is part of what forms the truth of God's grace. Verse 19 says, "And what is the exceeding greatness of his power to us-ward who believe, according to the working of his mighty power." That's not a bad definition of grace: His power to us. The things that are over my head are under Christ's feet.

Everything you need today is found in Christ, and everything you have that's worth having is from Christ. Ephesians is full of examples of the privileges we have in Christ. In verse 6 we are told that *"he hath made us accepted in the beloved one."* The Bible says we are chosen *"in him"* and that we occupy a heavenly position *"in Christ."* It says that grace comes to us *"from the Lord Jesus Christ."* I don't know what you face today, but God does. Christ can, and He waits simply for you to ask for His help and obey.

◇◇◇

A PERSON, NOT A THING

DAY 305

Ephesians 2:14 For he is our peace, who hath made both one, and hath broken down the middle wall of partition between us;

Are your kids afraid of storms? I mean the kind of storms with lightning, thunder, and lots of rain! If they are, what do they want when they come to you? They want safety, security, and peace; but do they say, "I want security"? No. They often cry, "I want my mommy!" They don't know enough to ask for security, but they instinctively know that all the things they need are provided **in a person**—namely, Mommy.

I don't know what you need this morning. Maybe you need safety, security, and peace; or maybe you need wisdom, money, and favor with people. I've asked for all of those things at one time or another. None of those things are wrong, but I can tell you that they are all found in a Person—Jesus Christ. **If you have Christ, you have everything you need.**

Jesus Christ can bring peace between a person and God or between two people, but notice what the Bible says: *"For he is our peace...."* Not only can He give and provide peace, but He IS peace. In other words, if you need peace this morning, you do not need a gift from Jesus in the form of peace; you need the Giver, Christ, Who is peace.

As a Christian, I can ask God for anything because of Christ's merits, not my own. Often we can come to our Heavenly Father in prayer and almost apologize for asking, and we tell Him how unworthy we are. Guess what? We will never be worthy, no matter how long we live! Any right you have to ask God for things is only because of Jesus Christ's merit. What is His is mine, for eternity and for today. As a child of God, I am favored by God Almighty, but not because of anything I have done. **God's favor is solely based on His grace and Christ's merits.**

Jesus Christ has what we need, knows what we need, and is able to give what we need! More than that, He IS what we need. What you need is found in a Person, not in a thing. No matter what the need is, go to the Person Who can provide what you need today!

◇◇◇

DAY 306 — HOW GOOD IS OUR UNITY?

Ephesians 2:19 Now therefore ye are no more strangers and foreigners, but fellowcitizens with the saints, and of the household of God.

Droid or iPhone? I imagine you have an opinion! Honestly, I could slice and dice everyone into many different groups based on the type of phone you prefer. It is amazing what we depend upon for a sense of unity. We actually have an affinity for people who use the same type of phone we use—an iPhone, a Droid, or something else. While those things are not necessarily evil, the fact is that **our unity is only as good as the one who unifies us.**

In the book of Ephesians, there are three "fellows" worthy of consideration. The first one is found in Ephesians 2:19 where Paul says, *"Now therefore ye are no more strangers and foreigners, but fellowcitizens with the saints, and of the household of God."* He is talking about both Jews and Gentiles, two distinct groups that, before Christ, would have been cut off one from the other. But he says in Ephesians 2:13, *"But now in Christ Jesus ye who sometimes were far off are made nigh by the blood of Christ."* When we draw close to Christ, we draw close to those who belong to Christ. We are fellowcitizens of the kingdom of God.

Ephesians 3:6 says, *"That the Gentiles should be fellowheirs, and of the same body, and partakers of his promise in Christ by the gospel."* So, we are fellowheirs. We have an inheritance given to us by God and provided to us by Christ. There is enough to go around! God's grace is not given to those who stand first in line or stiff-arm their way through life. God's grace is given to the humble, and there is more than enough for everyone who is in God's family. We are fellowcitizens, and we are fellowheirs.

In Ephesians 3:9 we have fellowship. It says, *"And to make all men see what is the fellowship of the mystery* [that Jews and Gentiles are saved by faith in Jesus Christ], *which from the beginning of the world hath been hid in God, who created all things by Jesus Christ."* There are some people for whom I just have an affinity. I do love people who are skiers or hikers, who love horses or mountains, or who like certain kinds of devices, but the bottom line is that the greatest fellowship is the fellowship we find in the Lord Jesus Christ because our unity is only as good as the one who unifies us.

DAY 307

ACKNOWLEDGE YOUR PART

Ephesians 4:7 But unto every one of us is given grace according to the measure of the gift of Christ.

Have you ever seen a caricature of yourself? It's a funny portrait, is it not? Humor is truth exaggerated, and that is exactly what a caricature is. The difference between a caricature and a portrait is the difference between exaggeration and accuracy. A caricature takes some physical trait and exaggerates it. For instance, if I happen to have slightly prominent ears, a caricature basically is an ear with legs; or if my hair has a certain kind of wave to it, the portrait will be nothing but hair. When it comes to a portrait of Christ, we need to portray Him with accuracy, unity, and health. We cannot settle for merely a caricature of the Lord Jesus by drawing attention only to our own part of the "body."

Ephesians 4:7 says, *"But unto every one of us is given grace according to the measure of the gift of Christ."* The Lord Jesus has gifted us to edify or build His creation, the body. The body of Christ is a unit. It is made up of many parts. God has gifted us to have a part in His creation. That's why the Bible says in verse 4, *"There is one body, and one Spirit…one Lord, one faith, one baptism, one God and Father of all, who is above all, and through all, and in you all. But unto every one of us is given grace."*

Spiritual health comes when I acknowledge God's creation and my part. My focus is to be on the Giver more than the gift. God has gifted me, but God hasn't gifted me to build me. He has gifted me in order to build the body of Christ. I need to focus more on my giving than on my gifting. God has gifted me to bless others, not merely to put the focus on me.

Verse 16 says, *"From whom the whole body fitly joined together and compacted by that which every joint supplieth, according to the effectual working in the measure of every part, maketh increase of the body unto the edifying of itself in love."* God has gifted every one of us to serve Him and to build the body of Christ. Do your part to bring glory to Him, not to yourself.

WHO ARE YOU IMITATING?

DAY 308

Ephesians 5:1 Be ye therefore followers of God as dear children.

Isn't it amusing when kids copy their parents? If you ever see my sons, Wilson or Weston, you cannot help but know to whom they belong. Why is that? Because they look and act like their dad! On purpose or by default, my kids imitate me.

Ephesians 5:1 tells us that we are to be **imitators** ("followers") of our Heavenly Father. When we *"walk in love,"* as Christ loved us, we imitate God and He is pleased (verse 2). However, if you are honest, some people are loved by God but not by you, right? Are there people that you do not even like? That is human nature; but God's nature is to love. **The only way to love people is for God to love them *through* you.**

If I am imitating God, who sacrificially gave Himself for me, I will be long-suffering, not short-tempered. If I am imitating God, I will be humble, not puffed up. I love others by doing right by them. **I may not always like everybody, but I am commanded to always do right by everybody.**

My feelings about people are fickle and can change at the drop of a hat. We usually like or dislike someone based on how they treat us, what things we have in common, and how we feel about them. Can you see how easily those things can change? In contrast, doing right by people never changes, no matter how they treat us, how we feel, etc.

Will Rogers used to say, "I never met a man I didn't like." I guess I have met more people than that! I may not like everybody, but by God's power I am to love people. That means I am to love even those people that are hard to love. Furthermore, **the best place to start loving people is in your own home.** It would be wrong to talk about loving all the people in the world and not do right by those who are closest to you! Today, allow the Spirit of God to work through you as you imitate the Father.

DAY 309

REFLECTIVE RELATIONSHIPS

Ephesians 5:21 Submitting yourselves one to another in the fear of God.

How easy it would be to be spiritual if we didn't have to hang around other people! There is more than a little irony in what I just said. In the first place, God does not want us to act spiritual. He wants us to be Spirit-filled. Secondly, if I am not right when I'm around people, then I am not right with God. **Our relationships to other people are a reflection of our relationship to God.**

Ephesians 5 talks about the importance of being governed by God's Spirit. Ephesians 5:21 says, *"Submitting yourselves one to another in the fear of God."* It is all too easy to think that you love God, but just can't stand people. If I am not loving people and doing right by people, I don't love God. I am not doing right by God. It is easy to have an attitude that says, "Well, I obey God only. No one else tells me what to do." If I am not obeying the authorities God has placed in my life, I am not obeying God either because my religion is only as good as my actions.

Ephesians 5 and 6 are full of examples of relationships between people, reflections of our relationship to God. Ephesians 5:22 says, *"Wives submit yourselves unto your own husbands, as unto the Lord."* Verse 25 says, *"Husbands, love your wives, even as Christ also loved the church and gave himself for it."* Ephesians 6:1 says, *"Children, obey your parents in the Lord: for this is right."* Verse 5 says, *"Servants, be obedient to them that are your masters according to the flesh…"* Lastly, verse 9 says, *"And, ye masters, do the same things unto them…knowing that your Master also is in heaven."* Our relationships to others are a reflection of our relationship to God.

WORK IN PROGRESS

DAY 310

Philippians 1:6 Being confident of this very thing, that he which hath begun a good work in you will perform it until the day of Jesus Christ:

For some time, there was a vacant lot where a church was going to build a new auditorium. A big sign on the lot said, "Look what God is doing here!" I thought that was a great idea for a while, but eventually I was embarrassed for the Lord and these dear folks. After months of looking at the unkempt and overgrown lot with no noticeable progress, one got the idea that God was not doing much there!

Sometimes I wonder if the same embarrassment would come if people had signs like that over their heads that said, "Look what God is doing here!" Would your life be a testament to God's working, or would your life be a little embarrassing by the lack of progress? The believers at Philippi were faced with the same question when Paul encourages them that *"he which hath begun a good work in you will perform it until the day of Jesus Christ:"*

You can live in confidence today by remembering that God has started a good work in you, and He *will* perform it until the Lord returns. **You are a work in progress!** It is true that God is still working on us all; **no one has arrived!** No matter how old you are or how much you are supposed to "have it together," you are still a work in progress. God will complete what He started!

Since God is still at work, we should not despair over our own lives or the lives of those with whom we work. Isn't it easy to be discouraged when working with kids or teens (or adults, for that matter)? Thank the Lord for opportunity you've been given, and get to work! There may be reasons to despair, but you can take comfort in the fact that **God *will* finish what He started!**

We can have complete confidence in God's work. Today, others ought see the good work that God is performing in your life. **So let's get to work—God is!**

DAY 311

WHAT AND WHY

Philippians 1:27 Only let your conversation be as it becometh the gospel of Christ: that whether I come and see you, or else be absent, I may hear of your affairs, that ye stand fast in one spirit, with one mind striving together for the faith of the gospel.

Philippians is a book that focuses our attention on **what we do** and **why we do it**. For Paul, the *what* and the *why* were the same. What he was doing was anchored in the gospel; why he was doing it was anchored in the gospel. In ministry, in your church, and in your own life spiritually, you may do everything efficiently, correctly, and well. But you are a failure if God is not in it, and you are not *"striving together for"* the same gospel.

"Striving" is not good if it is not *"striving together."* *"Striving together"* is not worthwhile if it is not *"striving together for."* The word *together* speaks of the people, and the word *for* speaks of the purpose. If you are missing either element, you really are missing out.

What should drive our *what* and our *why* is the gospel. You will find the word *gospel* throughout Philippians 1, and I think that is very instructive. The gospel affected Paul's attitude and his fellowship with others (see verses 5, 12, 18, 27). Paul was optimistic, selfless, confident, and content because of the gospel.

I recently took a trip to the Philippines, and I had the privilege of preaching to a group of pastors. Although I had never met any of the Filipino pastors, I had sweet fellowship with them during my visit. In a country with a different culture, different politics, and numerous other things, our fellowship was not based on the superficial. The fellowship we had was good because of the gospel of Jesus Christ.

What you do and why you do it ought to be wrapped in the gospel. The gospel will help you have the right focus as you are *"striving together for...."* Life is full of conflict, but a gospel focus will determine the people I work with and the purpose of what

I do. **Harmony is not worth much at all if it is harmony in the insignificant or inferior**. Strive together for the faith of the gospel. Take a moment to consider your *what* and *why* for today.

YOU IN FIVE WORDS

DAY 312

Philippians 2:25 Yet I supposed it necessary to send to you Epaphroditus, my brother, and companion in labour, and fellowsoldier, but your messenger, and he that ministered to my wants.

If you had five words to describe yourself, what would you say? I have actually done this recently, and it is very intriguing! It may seem difficult to describe yourself in five words, but that is about all a person gets on a tombstone. So if you had to pick five words to describe yourself, what would they be? Whether you like it or not, there *are* five words to describe you!

Most of the words we would think of probably have both a positive and a negative connotation. For example, if I described someone as "determined," that sounds good. However, "determined" could mean "stubborn" and "hard-nosed" too! God has made each of us different, and we are all gifted in different ways. Do your five words honestly describe the type of servant God wants?

Paul describes Epaphroditus five times in verse 25: *"brother ... companion in labour... fellowsoldier ... messenger ... he that ministered...."* The relationship, the labor, the striving, the message, and the ministering were all based upon the gospel. The book of Philippians is full of exemplary people. Paul, Epaphroditus, and Timothy are just a few examples found in Philippians 2, **but the ultimate Example is found in verses 5-8**: *"Let this mind be in you, which was also in Christ Jesus."*

Take a few minutes this morning to honestly evaluate yourself in five words. Don't think of what you *want* to be; but think of what you *are* right now. If you find that your five words are not what would please the Lord, ask His help today to go about changing them!

DAY 313

NO CONFIDENCE

Philippians 3:3 For we are the circumcision, which worship God in the spirit, and rejoice in Christ Jesus, and have no confidence in the flesh.

It is amazing what people do from the innate desire to have peace and know God. On a recent trip to the Philippines, I saw news coverage of a statue that dates back hundreds of years which is worshipped by huge masses of people. The desire to have peace, to know God, and to do something to gain God's favor spans time and culture. Paul didn't worship this statue from the Philippines; but before his conversion, all his confidence and boasting was in what he had done, not in Who Christ is.

Is it possible to be guilty of the same thinking, even after we are saved? Any person on any given day can transfer his confidence from Christ to himself, in Christian service or any number of areas in the Christian life. One sure-fire way to know in whom you've placed your confidence is to observe what you rejoice in. Paul challenges us to *"rejoice in Christ Jesus, and have no confidence in the flesh."*

What do you have to do today? Where is your confidence? If your confidence is in *yourself* now and in what *you* have done later, you are carrying a heavy weight! That kind of confidence depends on your goodness, and the odds are not stacked in your favor on that one, either! **If today's success is based on *your* strength, you are carrying a burden that does not belong to you, and you are in big trouble.**

If your confidence is in Jesus Christ, *He* is the One who has to be strong. If you have trusted Christ for salvation, don't you think He is trustworthy enough to put your confidence in today? Wonderful confidence comes by trusting God. **Confidence in Christ is steady because He never fails!**

Everything Paul had going for him—everything *he* had done—was *"counted loss for Christ."* The currency he had stored up was not legal tender in heaven! Only *"the righteousness which is of God by faith"* is accepted by God. If you have trusted God for salvation, there is no need to carry the burden that comes along with confidence in yourself. **There is no confidence like confidence in Christ!**

"... rejoice in Christ Jesus, and have no confidence in the flesh."

◇◇◇

AT HAND

DAY 314

Philippians 4:5 Let your moderation be known unto all men. The Lord is at hand.

Do you know people that are "mood changers"? The moment they walk into a room, the mood changes! Have you ever had a conversation about someone as he entered the room? Instantly, the mood changes (for a different reason). Philippians 4:5 points us to an experience like that: *"Let your moderation be known unto all men. The Lord is at hand."*

We are supposed to let our "moderation" (gentleness, sweet reasonableness) be known unto all men. People should "pick up" on our gentleness because *"the Lord is at hand"*—God is right here, at arm's length. "At hand" could mean closeness in time (He is coming back soon) or proximity (He is literally right here). Both are true, and the rationale is the same either way. Someone "at hand" is right *here*!

This business of gentleness ("moderation") is not a matter of weakness; it is a matter of meekness. **The fact that *"the Lord is at hand"* produces for us a gentleness because of His power.** Let me illustrate. In the Philippines, I saw several armed guards at the stores in the mall. Not only were they outside every store, they were all carrying 12-gauge shotguns! I never observed a guard in a panicked or threatened state. He could afford to be cool, calm, and "gentle" because of the firepower he had over his shoulder! As believers, we are to show meekness because the Lord and His power are at hand.

Would people know that God is "at hand" by the way you act today? I began praying last summer that I would literally know God's presence in counselor devotions, in my afternoon meetings, and all throughout my day. His presence was there; I just wanted to be aware of it. What I found is that when I was pressured and my patience had worn thin, my natural response was always disappointing. However, when I recognized that God was *"at hand,"* that made all the difference! Don't you think God's power is capable for each situation of your day?

The fact that God is "at hand" can give you calmness, comfort, and confidence for your day. If another human being can change the mood by walking into a room, how much more should the truth that *"the Lord is at hand"* change your actions today. We would do well to live in light of Philippians 4:5 today.

"Let your moderation be known unto all men. The Lord is at hand."

◇◇◇

DAY 315
THE GOSPEL GOES WITH YOU!

Philippians 4:22 All the saints salute you, chiefly they that are of Caesar's household.

Recently on a Friday morning here at camp, I was addressing the adults that brought young people from literally all across the country and abroad. I was struck by the fact that within 48 hours, we would be spread across a distance that reached from Yuma, Arizona, to Nassau, Bahamas. In short, we had, condensed in a single room, people who would spread out over the better part of a continent. How amazing!

God intends that the gospel go wherever you do. Philippians 4:21 says, *"Salute every saint in Christ Jesus. The brethren which are with me greet you."* Verse 22 says, *"All the saints salute you, chiefly they that are of Caesar's household."* This is amazing. These "brethren" are brothers with Paul in God's family. The "saints" are children of God in God's family. "Caesar's household" refers to the people employed by the tyrant of the Roman Empire. The point is that God intends the gospel to go wherever you do, whether that is to the California border, to The Bahamas, to the Midwest, or to any other place God has sent you.

In Ephesians 6:19, Paul asked that his friends would pray for him to have boldness to open his mouth to make known the gospel where he was. He goes on to say in verse 20, *"For which I am an ambassador in bonds: that therein I may speak boldly, as I ought to speak."* Paul did not ask for an open prison door or another opportunity somewhere else. He just asked that where he was, which happened to be in prison, God would give him an open door to share the gospel.

I don't know where you live or what you do today, but I do know that regardless of the atmosphere in which you find yourself or the kind of people with whom you make contact, God intends for the gospel to go wherever you do. If you will ask God for an open door today, I believe He will give it. Paul had an open door in prison to give the gospel, and God will surely give you an open door to give the truth where you are today if you'll ask for it, seek it, and seize it.

THE POWER OF PATIENCE

DAY 316

Colossians 1:11 Strengthened with all might, according to his glorious power, unto all patience and longsuffering with joyfulness;

What kind of power impresses you? For some, it may be the power of the truck you drive. For others, it may be the power someone has exercised in the political arena. The truth is, there is more than one kind of power. Colossians 1 reminds us of a kind of power that we often overlook.

What would you change if you had all *"all might"*? I can tell you that I have a few things on my list that I would change!

Sometimes God gives us the power to change *self* instead of the power to change other things. That kind of power is significant!

Colossians 1:11 reminds us that **the power of God gives us the power to have patience.** If I had my choice to have either the power to change something or the power to be patient, I would not choose patience! However, our wise and perfect Heavenly Father wants to give us the power to be patient.

Sometimes, the most important power you can have at your job, in your classroom, or in your ministry is the power to be patient. The word *longsuffering* paints quite a picture, doesn't it? **It is one thing to be patient with someone; it is another thing to suffer long with that person!** Don't forget that sometimes the people who need the most help require the most patience. Patience doesn't come naturally for any of us; remember that *"all might"* comes from Almighty God!

Today, we need *"all might,"* not just to change things, but also to patiently endure things as God sees fit. Only God has the kind of power that can patiently endure and can do so with joy. None of us have that kind of power in ourselves, but I am thankful we have a Father who has *"all might."*

◇◇◇

DAY 317 — IS THE TRUTH WORKING AT HOME?

Colossians 3:17 And whatsoever ye do in word or deed, do all in the name of the Lord Jesus, giving thanks to God and the Father by him.

I've come to the conclusion that it is easier to be spiritual when you don't have kids. Now, I'm joking a little bit, but what I mean is that it is easier to put on a good appearance to others when you don't have children than it is when you have children who reflect your most unguarded words and actions at home. Thus, I have come to the conclusion that **truth that isn't working at home isn't working.** Now, the truth always works; but I'm simply saying that if I am not living the truth at home, I am not living the truth. I am not allowing God to have His way through my life.

Colossians 3:17 says, *"And whatsoever ye do in word or deed, do all in the name of the Lord Jesus, giving thanks to God and the Father by him."* This comes on the heels of admonitions from God to have mercy, kindness, humility, meekness, and patience. Verse 13 says, *"Forbearing one another, and forgiving one another."* God also tells us to have the peace of God in our hearts and, in verse 14, to *"put on charity, which is the bond of perfectness."* If there is anything we need in our home, it's the superglue of supernatural love, God's love through us.

Now all these virtues are things that we want, but we need to allow God to work these through us in our own homes, first of all. That is why this passage addresses wives in verse 18, husbands in verse 19, children in verse 20, and fathers in verse 21. The bottom line is that a religion that is not working at home is a religion that is not what it ought to be.

Our goal ought to be being loved and respected the most by the people that know us the best. I'd rather be loved and respected by my ten-year-old than be admired by some prominent person in a church somewhere, because my ten-year-old knows me better. But I can only do what I ought to do as I allow the Word of God to dwell in me, enrich my life, and spread from my home to the world at large.

◇◇◇

SERVING GOD OR MEN?

DAY 318

Colossians 3:22 Servants, obey in all things your masters according to the flesh; not with eyeservice, as menpleasers; but in singleness of heart, fearing God:

Do you care what people think? Sometimes a person will boastfully say, "I don't care what anyone thinks!" I'm sorry, but I don't buy that! And, if it were true, I don't think I would want to be around that person! All of us care what people think. The better question is, whose favor are you most concerned with cultivating?

You are not serving God if you are not serving men. God calls the authorities He has placed over you *"masters according to the flesh."* The authorities we are to obey have hides and bones! A rebel can easily talk of obeying someone that no one sees; but obeying the boss he can see is another matter. You are not obeying a God you cannot not see if you do not obey the authority you *can* see.

Likewise, **you are not serving God if you are *only* serving men.** The Bible continues by saying, *"Not with eyeservice, as menpleasers; but in singleness of heart, fearing God."* It is possible to please everyone around you, including your pastor, boss, and teachers, but not be right with God. I think every Christian probably wants to please God; but if my work comes down to *"eyeservice"* and pleasing men only, something is wrong.

When you boil it all down, you serve the Lord. You cannot serve God without serving men, and you cannot serve God by only serving men. Today, whose favor will you strive to gain?

"Servants, obey in all things your masters according to the flesh … and whatsoever ye do, do it heartily, as to the Lord, and not unto men."

DAY 319
AUTHORITY: STEWARDSHIP AND RESPONSIBILITY

Colossians 4:1 Masters, give unto your servants that which is just and equal; knowing that ye also have a Master in heaven.

When I was 13, I kept waiting for my growth spurt to occur. People told me, "When you're fourteen, you'll probably grow a couple of inches. I waited, and it never quite happened. Then it was supposed to happen at fifteen, sixteen, seventeen, and so on. Today, I'm a full grown man who really never grew to the height he had anticipated. Maybe that's the way you feel about the growth of your authority. When you're 13 you long for the day when you will be the boss and call the shots. You think that will happen at 18, and it doesn't. And then you think it will happen at 21, and it doesn't. Now, someone else is still calling the shots. They are telling you when to show up, what to put out, when to hang it up, when to go home, and you are thinking, "Will I ever be in charge?" No one who is not under authority has authority.

In Colossians 3, the Bible uses the words *submit* and *obey* when it is talking about wives, children, and servants. Maybe you think that you should be in charge. Well, Colossians 4:1 says, *"Masters, give unto your servants that which is just and equal; knowing that ye also have a Master in heaven."* The Bible tells us in the previous verse that there is no respect of persons with God. No one is going to buy or bribe God's favor, nor will they bribe or beg for authority from God. There is no respect of persons.

Your responsibility to your children or to your subordinates at work is framed by your responsibility to God. *"Masters, give unto your servants that which is just and equal; knowing that ye also have a Master in heaven."* No one who is not under authority has authority. Show me someone who thinks that they do not answer to some authority, and I'll show you a person who is a menace. Whether you are a child or a father, a preacher or a layperson, each one of us is under authority and honors God with a spirit of submission.

While the text here does not speak of it, many other places in the Bible are very clear about the fact that all of us have human authority. There is but one Authority in this universe, and that is God. Thus, all authority (President, preacher, father, etc.) is delegated authority. When I cut myself off from those who are in authority over me, I cut myself off from power to lead those who come behind me under my authority. The bottom line is that **authority is nothing more than a stewardship and nothing less than a responsibility.** If I have authority, it is a stewardship that belongs to God and is entrusted to me. If I have authority, it is a responsibility to which I will answer to my authority, God.

DAY 320

STEWARDSHIP

I Thessalonians 2:4 But as we were allowed of God to be put in trust with the gospel, even so we speak; not as pleasing men, but God, which trieth our hearts.

In this passage, we see that Paul's main concern in giving out the gospel was his testimony (I Thessalonians 2:3-13). He was interested in *how* he spoke, not just in *what* he spoke. The gospel was the message; Paul was simply the messenger. But how did Paul view the gospel?

Verse 4 says, *"But as we were allowed of God to* **be put in trust** *with the gospel...."* Verse 13 adds, *"... when ye received the word of God which ye heard of us, ye received it* **not as the word of men**, *but as it is in truth,* **the word of God**...." What Paul had—the gospel—belonged to God! That truth is what fueled his concern about his testimony in verses 3-13!

Recently, I was preparing for a trip to the West, and it occurred to me how much *one* church was investing in one short week. While the trip included airfare, lodging, meals, and time, it was all for the purpose of preaching the Bible to adults and young people. In a sense, I was a steward of what this church entrusted to me for the week!

The same thing is true with what you have and what you do every day. Everything you have belongs to God. Everything you have is a stewardship. This means that laziness, wastefulness, and bad attitudes are a serious offense to the One who gave everything to us!

The Ranch is not mine; it is the Lord's. The buildings, the horses, the equipment—it all belongs to Him. That makes what I do with them mightily important! The reason God has given all these things is for the most important stewardship we have—the gospel. How are you doing with the stewardship you've been given?

WILD MUSTANGS

DAY 321

I Thessalonians 2:19 For what is our hope, or joy, or crown of rejoicing? Are not even ye in the presence of our Lord Jesus Christ at his coming?

Miller Coliseum in Murfreesboro, Tennessee, is the largest building of its kind in our area. Rodeos, horse clinics, and all types of equestrian events are held there. In October, an event called "Mustang Makeover" was held at the coliseum. The idea is to take a wild mustang, break him, and then show other cowboys your handiwork at the makeover event!

Sometimes camp (or your church youth ministry) can feel like "Mustang Makeover." Are you thinking of a particular teenager when you think of breaking a wild horse? Some "wild" teens never grow up; and years later, pastors and Sunday school teachers are left to work with "unbroken" adults. Just like horses, **we need to have our wills broken.** We ought to allow the "rein" (reign) of God to rule in our lives.

In ministry, oftentimes, we can lose sight of what really is important. If you live for and expect a big payday, large numbers, or even recognition and thankfulness from those you serve, you are destined for disappointment. I Thessalonians 2 tells us what is truly important in life. Paul reminds us that the very challenge of ministry is also the reward! **Problems today can be your prize tomorrow.** The "wild mustangs" you are breaking today can one day be your *"hope, or joy, or crown of rejoicing."*

Today, remember that the greater your burden is now, the greater your joy can be in the future. When you have problems, do not get so focused on breaking mustangs that you lose sight of the reward coming later! This truth also applies to raising kids or ministering in your church. The prize-winning cowboys at the "Mustang Makeover" aren't winners because they broke old, haggard horses; they won because they broke spirited, strong-willed horses. The reward comes *after* all the hard work!

Only God can change hearts; and only God can empower you to make a difference to those you serve. **Remember that the greater the burden is *now*, the greater the reward will be *later*!** Rely on God's power this week as you work with people who are a bit "wild" and "unbroken." Let's saddle up and get to work!

◇◇◇

DAY 322 — GOD'S WILL IS HOLINESS

I Thessalonians 4:7 For God hath not called us unto uncleanness, but unto holiness.

What is God's will for your life? What is God's calling on your life? Someone may say, "Well, Brother Wil, God has called me to preach." Someone else may just as easily say, "Well, Wil, I am working with computer programming, but I can't say that it is God's calling. I just don't know." I can tell you what God's calling is. Now I'm not presuming to know the details of God's plan for your life, but I do know part of God's calling because I Thessalonians 4:7 tells us, *"For God hath not called us unto uncleanness, but unto holiness."* So as a believer in the Lord Jesus, God has called you to holiness—to be separate, to be different. That is God's call.

Holiness is more than just what you do. It is what you are. You are accepted in Christ. If you are saved, you are fundamentally different from everybody else. You are not going to hell; you are going to heaven. You are not part of darkness; you are part of light. You are in Christ. Your actions need to reflect and be in harmony with what you are. You are a work of Christ.

What is God's will? I Thessalonians 4:3 says, *"For this is the will of God, even your sanctification, that ye should abstain from fornication."* God's will is more about your integrity and your life matching your position in Christ than it is about your vocation or where you go. I am not saying that God doesn't have specific details lined out for you. I am simply saying that those don't matter if you are not taking care of the most basic areas of your life.

Whatever else God wants you to do, He wants you to be set apart to Him, not only in your position but also in your actions. We are holy, and we are to be holy. God is most concerned about what you are at heart when no one is watching. There should be a level of integrity that's a platform for you to do whatever God has gifted you to do. God's will for you today as a child of God is to live holy and to do right. Your actions should be a reflection of your position in Christ.

OUR FAMILY, THE FAMILY OF GOD

DAY 323

I Thessalonians 5:11 *Wherefore comfort yourselves together, and edify one another, even as also ye do.*

The largest family I know could not fit in your minivan. In fact, it couldn't fit into two minivans, a bus, or a parking lot. I'm talking about the family of God! Recently, I noticed how frequently the Bible refers to family relationships within the family of God. In I Thessalonians 5 specifically, you find words like *brethren*, *brothers*, and *children*.

I Thessalonians 5:4 begins with *"ye, brethren."* Again, in verses 12, 14-15, and 25-27, the Bible talks about brethren. Verse 5 says, *"Ye are all the children of light,"* speaking of the children of God.

Your Christian testimony needs to begin in your house; but don't neglect the larger family, the family of God, that can help your house. I do believe that my home is of primary importance in my life; but I cannot honestly say that my family comes first if I am neglecting the family of God, because my church family helps my home. God has put me in this family, and the help I need comes from this family's fellowship, instruction, and teaching. Church is important. **Let the family of God edify the family in your house.**

Additionally, your family should edify the family of God. I Thessalonians 5:12 says, *"And we beseech you brethren, to know them which labor among you, and are over you in the Lord, and admonish you."* Paul is talking about people that teach, lead, and guide. The next verse says, *"And to esteem them very highly in love for their work's sake. And be at peace among yourselves."* We are to support, honor, and love those that guide us.

We are also to support one another. Verse 14 says, *"Now we exhort you, brethren, warn them that are unruly."* He is talking about the help we are to be to the family of God. Verse 25 says, *"Brethren, pray for us."* It may be a scary world, but it does not have to be a lonely one. God has given you His Spirit, His Word, and a church that can help and encourage you while you do the same for others.

DAY 324

RESPONSIBILITY AND ABILITY

I Thessalonians 5:14 Now we exhort you, brethren, warn them that are unruly, comfort the feebleminded, support the weak, be patient toward all men.

Our world contains such a wide variety of people! We are all "bent" in unique ways. Some are naturally prone to give comfort; others naturally want to whack people over the head! God has gifted each of us in particular ways and has given us different strengths. It is important to remember that **the type of your service depends on the needs of those you serve, not on your own abilities.** In other words, your "job" is more about *responsibility* than *ability*.

What happens when the need exceeds your abilities? If you are relying solely on your abilities, you are up a creek without a paddle! However, if your responsibility is God-given, and your gifts are too, then you have everything you need! God's power is supernatural, and it can rise above the greatest need you will face.

How can you warn, comfort, and support, while having the wisdom to know what is needed? Only God can give you that! Some people you serve need to be warned because they are unruly; some need comfort because they are feebleminded; and some need support because they are weak. On top of that, God says we ought to have patience toward all men. That's a tall order! You may not have one of those gifts, or the patience you need for every person today; but **if God has given you the responsibility, rely on His ability!**

Different people have different needs that require different actions. You should not live your life saying, "Well, that's not my gift!" It is not about your ability! Ask God to give you what you need. We all have certain strengths and weaknesses, but we all need the supernatural power of God to serve those around us, whatever their needs. Instead of focusing on your ability today, focus on the responsibility God has given you in your church, home, and workplace. Ask Him for what you need!

◇◇◇

IS EXERCISE GOOD?

DAY 325

II Thessalonians 1:4 *So that we ourselves glory in you in the churches of God for your patience and faith in all your persecutions and tribulations that ye endure.*

Do you have a plan for better health this next year? You may be planning to run, to lift weights, or to exercise a set amount of time each day; but whatever you do, improving your physical health requires resistance. Resistance doesn't feel good to your body, but straining your muscles in a good way helps to strengthen them. The same is true of Christian virtues that should grow in your life. **The good things that God wants to build in your life will come with some resistance and straining!**

Paul revealed his thankfulness for the Christians at Thessalonica when he said to them, *"Your faith groweth exceedingly, and the charity of every one of you all toward each other aboundeth."* How did this all come about? Through enduring persecutions and tribulations with patience and faith! Their faith was growing because their faith was being resisted.

Do you thank God for an easy lot in life, or do you thank Him for the faith, patience, and love that grow when things aren't easy? Like me, you probably enjoy calm, ease, and peace; but none of these build virtues. **Virtues like faith, patience, and love grow in your life during the absence of "peace and quiet," rest, and easy times.**

No matter what your exercise regime, there will be some resistance. That is the way your muscles are strengthened and your body is benefited. In the same way, virtues like patience, love, and faith will grow when everything is not smooth sailing. Be thankful when God "exercises" you in order to grow you!

◇◇◇

DAY 326

FINALLY

II Thessalonians 3:1-2 Finally, brethren, pray for us, that the word of the Lord may have free course, and be glorified, even as it is with you: And that we may be delivered from unreasonable and wicked men: for all men have not faith.

Every work week has a Friday; every month has a last day. Profound, isn't it? There is finality in many of the things we often overlook in our daily routine. Even in a week of revival meetings or a week of camp, there is always a final service. The first part of II Thessalonians is timely for the end of a week, a month, and a year!

Paul made two appeals in II Thessalonians 3:1-3. First, he said, *"Brethren, pray for us…."* That is a pretty good request! I hope you will think to pray for us here at the Ranch; and you can certainly pray for your pastor, your church's missionaries, and other good Christian ministries. I have been praying for people, for businesses, and for churches this year precisely because they were praying for me!

What could you pray for? **Pray for power** (*"that the word of the Lord may have free course, and be glorified"*) **and for protection** (*"that we may be delivered from unreasonable and wicked men"*). Not everyone we work with is reasonable and nice; sometimes they are just plain mean! Yet the ministry God has called us to (and called your pastor to) demands that we sometimes work with *"unreasonable and wicked men."*

Paul's second appeal is to keep at it. The Bible encourages us by saying, *"But the Lord is faithful, who shall stablish you, and keep you from evil."* Because the Lord is faithful, we should be faithful too! **Anytime I am less than faithful, I am not following the pattern that God gave.** The Lord is faithful and just to forgive, and He is faithful to empower. He will give you exactly what you need in order to do exactly what He has called you to do! **Just keep at it!**

I could not say it any better than the way the Apostle Paul said it in II Thessalonians: *"Finally, brethren, pray for us… the Lord is faithful, who shall stablish you…."* At the end of another week and another year, what a great reminder to pray for others and to keep at it!

◇◇◇

YOUR RIGHTS

DAY 327

II Thessalonians 3:9 *Not because we have not power, but to make ourselves an ensample unto you to follow us.*

Do you know your rights? "I know my rights!" is a phrase you often hear in our day. People can become so wrapped up in their rights that they fail to examine whether something is actually right or wrong. You can do many wrong things with your legal rights! The truth is, **you can either follow your**

rights or you can lead by example. Those are your two choices, and you cannot do both at the same time.

Some Christians live with the phrase "chapter and verse" on the tip of their tongues. Whenever there is a question about application of Bible truth, the first thing on their minds and out of their mouths is, "Show me chapter and verse!" Of course, the Bible must be the authority for everything we believe and practice. However, a person with a sour attitude is not going to see the truth, even when you show them the chapter and verse. Forget Holy Spirit guidance, they just want a list of do's and don'ts. That is a pretty shallow way to live!

Always demanding your rights will not help you influence and help other people. **When you follow your rights, you lose your right to be followed!** You must decide to either be focused on your rights or to be focused on what will help others to follow the Lord. The best way to make sure others follow you is to do right yourself.

What is the governing ethic of the way you feel and think today? You may have the right to do something, but that doesn't mean it is always right! Could everyone follow your lead? Remember, you can either follow your rights or you can lead by example and influence.

◇◇◇

DAY 328
HEY, ARE YOU BUSY?

II Thessalonians 3:10-11 For even when we were with you, this we commanded you, that if any would not work, neither should he eat. For we hear that there are some which walk among you disorderly, working not at all, but are busybodies.

Is there a difference between "work" and "busyness"? In reality, this question of working and being busy is only a concern in our world. In the animal kingdom, there is no such thing as a lazy squirrel. Lazy squirrels are dead squirrels! On the other hand, have you ever seen a squirrel that was busy scampering back and forth in your lane of the highway? He just needs to make up his mind and get out of the way! He is plenty busy but is only accelerating the trauma he will meet when your tires roll around.

In II Thessalonians, some people were not working with their hands, but they stayed busy. Paul called them *"busybodies."* You can be very busy inserting yourself into someone else's business, but it is another thing altogether to be industrious. **Busy speaks of energy;** *work* **(industry) speaks of an end result.**

You should focus your day on what you accomplish, not on how busy you are. Focusing on filling time instead of accomplishing things will make you busy but not industrious. Work can be virtuous. Hard work is virtuous. But **the best work is important because of what it accomplishes.**

Make your day count for something. Be more concerned with the end results than with busyness. **The benchmark for today is not how busy you are but how much good you accomplish.**

THE GREATEST ABILITY

DAY 329

I Timothy 1:11-12 According to the glorious gospel of the blessed God, which was committed to my trust. And I thank Christ Jesus our Lord, who hath enabled me, for that he counted me faithful, putting me into the ministry.

If you could improve yourself in any way, what ability or skill would you choose? Would you pick improving your knowledge and use of computers, your ability to communicate to others, your physical ability to accomplish a task? I suppose we could all use a little improving! Today we are surrounded with books, classes, and resources to help us improve ourselves, but what does God consider important? We don't have to wonder; He tells us in I Timothy 1!

Our culture is enamored with abilities that are visible and get attention. There is nothing wrong with having an obvious ability, but what is the most important ability to God? Paul said that

God had committed [the glorious gospel] to his trust. Not only that, God had *"counted* [him] *faithful, putting* [him] *into the ministry."* No matter your abilities, the question is not, "Can I trust God?" The question is, "Can God trust me?"

"The greatest ability is dependability," as Dr. Bob Jones, Sr., used to say. Can you be depended upon? Can you be counted as faithful? Can God commit things to your stewardship? The question is less about *ability* and more about *dependability*. God will enable you when you are faithful to Him.

No matter what weakness you sense in your life, **you can always be faithful**. All things being equal, your church doesn't need more skill; your church needs faithful people. Your family doesn't need "super" parents; your family needs parents who will faithfully follow God's plan from the Bible. No matter what you do today, the greatest ability you have is dependability! Faithfulness is the ability God considers most important.

◇◇◇

DAY 330

BROAD OR NARROW?

I Timothy 2:4-5 Who will have all men to be saved, and to come unto the knowledge of the truth. For there is one God, and one mediator between God and men, the man Christ Jesus;

Have you ever met someone who thought he was broad-minded because he thought like the majority? There is a big difference between being with the broad majority and being broad-minded! In fact, sometimes the more widely accepted your ideas, the more narrow-minded you actually are! Oftentimes, just going along with the majority is easier than understanding or defending what you believe.

You can see life clearly when you think as broadly and as narrowly as God does. God says that we are to pray *"for **all** men,"* and that He wants *"**all** men to be saved, and to come unto the knowledge of the truth."* On the other hand, there is *"**one** God,* *and **one** mediator between God and men, the **man** Christ Jesus."* "One" is narrow while "all" is broad!

The Lord Jesus is God in flesh and the *only* way to God, but it is His desire that *all*—not some or a select few—men be saved. When you are too broad with the way to heaven, or too narrow with who can receive God's gift, you are directly opposed to God's clear plan!

Let's start with the most basic instruction given in I Timothy 2. God says we are to pray *"for all men … for kings, and for all that are in authority."* Do you think you should pray for the President and other elected officials? Do you think it would make a difference? God also wants *"all men to be saved."* Be as broad-minded as God is. Pray for all men and share the gospel with all men. Prayer does work, and the gospel still saves men!

On the other hand, don't fall for the lie that there is more than one God or more than one way to God's heaven. Jesus Christ is *"the way, the truth, and the life."* There is *"none other name under heaven given among men, whereby we must be saved."* You cannot compromise God's way to salvation! Be broad where God is broad and narrow where God is narrow.

◇◇◇

TO BE OR TO DO

DAY 331

I Timothy 3:10 And let these also first be proved; then let them use the office of a deacon, being found blameless.

I landed a really big part in my kindergarten graduation play as one of Little Bow Peep's sheep. Now I didn't have a speaking part, but I had a big cotton ball outfit to make me look like one of Little Bow Peep's sheep! Somehow people failed to see the brilliance of my acting. All joking aside, many of us have a hard time distinguishing between the character we possess and the role we play. We see ourselves as little more than the part we play. The role that God has given you in life is important, but you are more than the job you have or the calling that God has given you.

In I Timothy 3, God is giving instructions to those who will serve in special ways in the church. He talks to both pastors and people in the church. In I Timothy 3:10, He says of deacons, *"And let these also first be proved, then let them use the office of a deacon, being found blameless."*

Now what I want you to see is that there is a difference between the office I "use" and the person that I am. I Timothy 3 is not about how gifted a person is; it is about the choices such a person makes and the character he possesses.

Verse 1 says, *"This is a true saying, If a man desire the office of a bishop, he desireth a good work."* Such a man is not to seek the honor; he is to seek the work. And it is honorable work. A bishop then must *be*. It doesn't say a bishop then must *do*, but that he then must *be*. The point is that God is more interested in the character you possess than the part you play. Now they aren't mutually exclusive. One is foundational to the other. The way I use the office that God had entrusted to me depends upon how seriously I take the character, integrity, and decisions of my life.

If all you do is view yourself as the sum and substance of your gifts or office, your calling or place, then you are missing out. It's not wrong to identify yourself with whatever God calls you to do, but you are made up of more than your vocation.

Any job I have is but a stewardship. The integrity I possess or lack is something for which I alone will be accountable to God. Realize that you will use your office best when you are the person God made you to be in daily areas of responsibility.

HAPPY HOMES, HAPPY CHURCHES

DAY 332

I Timothy 5:1-2 Rebuke not an elder, but intreat him as a father; and the younger men as brethren; The elder women as mothers; the younger as sisters, with all purity.

Have you ever wondered why cranky people at your church are so cranky? What are they always unhappy about? What about the children who are brats at your church? Do you know some selfish people at your church?

On the other hand, do you know people who always seem to brighten your day? Do you know families with well-behaved children? Do you know people who are generous at your church? What makes the difference? Many times, it is the home. I Timothy 5 reminds us that the church and the home are inseparably connected. **Your church will not be any stronger than the homes that make it up!**

I thank God for good churches, good institutions, good schools, and good camps, but the main responsibility for your home begins with *you*. I am thankful for the godly influence that people have on my kids, but raising my kids is not the responsibility of my pastor, my child's teacher, or good camp counselors.

Paul tells Timothy to treat the older men in the church as "fathers," the younger men as "brothers," the older women as "mothers," and the younger women as "sisters." You can't treat people right in your church if you don't know how you are supposed to treat dads, moms, brothers, and sisters! If you get and give help at home, you will be a help at church.

The truth is, my kids will not do right by my pastor at church if they are not doing right by me at home. The same goes for you. **The place to learn about treating others properly is in your house.** How should you treat your pastor, a widow in your church, or young ladies in your church? Practice on your dad, mom, and sister!

Those people who learn well at home will live right at church. **The best way to be faithful at your church and in your ministry is**

to faithfully lead and obey in your home. If you will do right at home, you can be a help to the cranky people, the bratty kids, and the selfish people at your church, too!

DAY 333
SETTING PRIORITIES: TAKING RESPONSIBILITY

I Timothy 5:16 If any man or woman that believeth have widows, let them relieve them, and let not the church be charged; that it may relieve them that are widows indeed.

This past year I read about half of a book that had a premise with which I didn't agree, but there was one thing the author said that caught my attention. Essentially, he said that a lack of time is a lack of priorities. In other words, you always have enough time to do what is really important. It is easy to buy myself an excuse to not do what I should do by saying, "I just don't have the time." A lack of time equals a lack of proper priorities, but setting proper priorities means taking responsibility.

I Timothy 5 is all about taking responsibility, making priorities, and realizing how taking care of priorities helps us with other areas of life. I Timothy 5:16 says, *"If any man or woman that believeth have widows, let them relieve them, and let not the church be charged; that it may relieve them that are widows indeed."* This is a reminder that my family is my responsibility.

Now I thank God for my church, but God did not call my church to take care of my family in the same way that I, as the father, am obligated to do. Because God has called me to care for my family, my church is very important because I need that help. But I am responsible to take care of the people closest to me in life. The church can help, but no one should take care of that which should be a priority to me.

I Timothy 5:8 says, *"But if any provide not for his own, and specially for those of his own house, he hath denied the faith, and is worse than an infidel."* If I am not making a point to take care of what is most important, I'm not honoring God.

Here's the bottom line: **setting proper priorities means taking responsibility.** We need God's wisdom to do that because sometimes we don't know where to start or what is most important. Do not allow this idea that you can't do all that you want to do in a day be an excuse for not doing what you should do. At the end of the day, sometimes we avoid setting proper priorities because we know on some level that setting priorities means taking responsibility. If you *should* do something, you *can* do something if you will trust God to help.

◇◇◇

WHAT WILL CATCH UP TO YOU?

DAY 334

I Timothy 5:24-25 Some men's sins are open beforehand, going before to judgment; and some men they follow after. Likewise also the good works of some are manifest beforehand; and they that are otherwise cannot be hid.

Can you imagine electing a President without ever watching a debate, knowing the issues, or considering what he thinks or believes? What if your church called a pastor without any regard to his beliefs, his philosophy, or his character? That is a recipe for disaster!

On the other hand, what happens when you think you have done your homework on a person, but you find out that you missed something? People may miss details while deliberating, but **God is never duped.** The Bible says, *"Some men's sins are open beforehand, going before to judgment; and some men they follow after."* Some people's sins are so obvious that they precede and announce themselves. Their reputation goes before them, and everybody knows what they are. Other people's sins are like cougars trailing their every step and waiting to pounce. These people go to bed every night with a smitten conscience.

In the same way, there are two types of good works: those that are *"manifest"* and those that are hidden (for now). **Some are** obvious; some are hidden; but both evil and good will catch up to you!** It is impossible to hide evil works, and it is impossible to hide good works. "Well, Brother Wil," you are saying, "I've done good things, and no one at my church even noticed!" Your good works may be hidden to your pastor, your boss, and your neighbors, but they are not hidden to God!

God is not mocked. A man will reap what he sows. Some men's sins are blatant and open, while other men's sins are trailing them like a hungry animal. Some men's good works precede them, and *"they that are otherwise cannot be hid."* Nothing is hidden from God. As one preacher put it, "There is a payday someday." So what is catching up to you today?

◇◇◇

DAY 335
FLEE AND FOLLOW

I Timothy 6:11 But thou, O man of God, flee these things; and follow after righteousness, godliness, faith, love, patience, meekness.

Have you ever taken the wrong road? **Did you know that it is impossible to take the wrong road and end up in the right place?** That's true, even if your intentions are good.

I Timothy 6:11 reminds us that you cannot flee and follow the same thing. Every day you make choices to either please yourself or follow God and what is important. **What you choose to follow influences what you flee.**

Elsewhere in this chapter, the Bible tells us that God has given us *"richly all things to enjoy."* Money is not wrong, and having nice things is not wrong. The trap and temptation is to allow the love of money (verse 10) and discontentment (verses 6-8) to ruin what God intended. The things we enjoy in life are gifts from God, but those gifts are not what we should pursue. If I expect to end up at the right place today (righteousness, godliness, faith, love, etc.), but I'm going down the wrong road (pleasing myself), I'll never end up at the right destination!

Do the words *righteousness, godliness, faith, love, patience,* and *meekness* describe your life? If you are not pursuing them, you will not possess them! If you are pursuing the path of least resistance or the path of pleasing self, you cannot possibly end up at the right place. All throughout today, you will have the choice to follow what your flesh desires or what God desires for you. What are you fleeing and what are you following? **Don't spend your day going down the wrong road!**

CONTENTMENT IN GOD

DAY 336

I Timothy 6:17 Charge them that are rich in this world, that they be not highminded, nor trust in uncertain riches, but in the living God, who giveth us richly all things to enjoy.

How much worry would completely melt away if you received a large sum of cash this morning? What kinds of things that presently concern you would be fleeting memories? How would your outlook on life change?

God has given us many things to enjoy. The Bible tells us that *"every good gift and every perfect gift is from above, and cometh down from the Father of lights…."* (James 1:17) Seeing every legitimate pleasure in life as coming from God puts a whole new, wonderful complexion on life! **You can find contentment *from* God, but it is even better to find contentment *in* God.**

Finding satisfaction in the good things that God gives is finding contentment from Him. Finding your worth and satisfaction in God Himself is even greater! I Timothy 6 reminds us that those *"that are rich in this world"* need to be warned against being highminded. Why would a rich man be highminded? He could find his confidence and self-worth in what he possesses.

Instead of trusting in *"uncertain riches,"* he should trust *"in the living God, who giveth us richly all things to enjoy."*

Nothing better describes this world's goods than *"uncertain riches."* Having money is not sinful; loving money is *"the root of all evil."* The word *uncertain* describes the stock market, the housing market, and just about anything else you can do with money! Instead of trusting in the good things that God gives, trust in God Himself.

Why is contentment in God better than contentment from God? **Everything that God gives and every pleasure that He allows is temporary and uncertain.** Nothing you can buy with paper money or a plastic card is everlasting. Contrast temporary things with the *"living God."* God will never run dry, run out, or be uncertain. He is the living God!

No matter how rich you consider yourself, there is a danger in seeking contentment from money. The best pleasures in life are the ones that God Himself has designed, but you ought not to find your confidence and self-worth in those good things. Instead, find your contentment *in* God.

DAY 337

TRUE FRIENDSHIP

II Timothy 1:16 The Lord give mercy unto the house of Onesiphorus; for he oft refreshed me, and was not ashamed of my chain.

The name Onesiphorus is probably not a very familiar one to you. However, it is a name that is found in connection with the Apostle Paul in II Timothy. When Paul was in prison, Oniesphorus was a friend to him. The Bible says he refreshed Paul. What that exactly means, I don't know. When I think of refreshing, I think of a cool drink, a cool breeze, a warm meal, or a time to relax. So his refreshing may have been just a kind word or some small token.

Whatever it was, it costed Onesiphorus something, probably a little bit of time or some money. Among other things, it at least costed effort. In other words, he was giving. Being Paul's friend wasn't "cool"; he did not do it for personal gain. He was Paul's friend because he had something to give Paul. He gave of his resources and probably gave of his reputation.

A few things come to mind from this friendship. First, **you discover who your friends are when you have nothing to give.** Many people come to your side when you're riding high and people recognize your ability to give to them, but you discover who your friends really are when you have nothing to give.

Secondly, you are being a true friend when there is nothing to receive from them but friendship. You are not demanding that they give you something; you are giving to them. That's what friendship is.

Thirdly, God knows. God knows when I have nothing to give and someone befriends me. God will bless them. And God knows when I am giving to others with no expectation of receiving. I'm not saying that friendship should always be a one-way street. I am saying that you and I can only make decisions for ourselves. We can make sure that we are givers.

Onesiphorus was a friend to Paul when it costed him. God knew that, remembered that, and blessed Onesiphorus for that.

BOUND OR NOT BOUND?

DAY 338

II Timothy 2:9 Wherein I suffer trouble, as an evil doer, even unto bonds; but the word of God is not bound.

Much was said some time ago about a Christian pastor in the Middle East who had been severely sentenced for his faith in Christ. This man, who is younger than I am, is being persecuted by his own people. When I think of those who try to squelch the gospel, I sometimes think of words like evil, unjust, and cruel, but II Timothy 2:9 reminds us of another word that describes the efforts of those who try to stop the Bible: futile.

Paul was in prison for spreading the gospel. Although he was in bonds, *"the word of God is not bound."* You cannot bind the Word of God, and any attempts to do so are simply futile. It's not going to work! You may imprison or execute God's servant, but you will not stop His Word and His power. Pilate and Caiaphas couldn't squelch God's Son, and no regime today will be able to exterminate the gospel.

The Bible is full of examples of evil people who tried to stop the power of God. Religious leaders wanted to kill Lazarus because of the publicity he gained after the Lord raised him from the dead. Haman tried to conspire to kill God's people, even building a gallows to hang Mordecai. Paul was imprisoned for preaching the gospel, but he just took the good news to jailers and Caesar's house! The early church was persecuted, and what happened? Did the gospel die because of terrible persecution, like the stoning of Stephen in Acts 7? No! *"Therefore they that were scattered abroad went everywhere preaching the word."* (Acts 8:4) The gospel grew precisely because of the persecution!

God's Word cannot be bound when God's people are obedient. So, the question for us today is, what has you bound? What is stopping you? There are people you should be influencing and people you should be winning to Christ. **Nothing is stopping you but *you*.** You can do what you ought to do. You and God make a majority. There is not a power on earth that can stop a Christian who is doing what God wants him to do! Are you living in obedience to what God has said?

"... but the word of God is not bound."

GOOD QUESTIONS

DAY 339

II Timothy 2:23 "But foolish and unlearned questions avoid, knowing that they do gender strifes."

Have your kids ever asked you a "good" question? A good question is often code for a question that doesn't have a good answer. Sometimes all you can say is, "That's a good question." As a parent, I've resorted to using that answer on several occasions! There are some questions that do not have good answers because they are smart questions. However, there are other questions that do not have good answers because they only incite contention. **A question is not a good question when, instead of providing direction, information, and answers, it only provides contention.**

Haven't we all been asked loaded questions that do not have good answers? Sometimes no matter how you answer, you are in trouble! Ironically, a question can be used to make a pointed statement. Is it any wonder the Bible warns us to avoid *"foolish and unlearned questions"*? **Avoiding these types of questions includes avoiding both asking them and answering them!**

One time a man raised his hand during the sermon I was preaching. I already knew the purpose of his question since he had previously expressed his anger with the previous night's message. He was lost and was not happy about what I was preaching. He disagreed with my notion that all men were sinners and that there was only one way to heaven. Right there on the second row, with anger and disgust, he raised his hand for everyone to see! I didn't take his question that day, but I can guarantee you he did not want an answer—he wanted contention!

Let me mention a couple of specifics that can help us all. First, watch out for complaining questions. A person can disguise a complaint as a question when he says, "Why does it have to be so hot?!" Does he *really* want to know the meteorological reason for the weather? No! He is, in fact, making a statement about the weather and complaining to someone who can do nothing about it! And he would not dare speak to the One who controls the weather in such a way! Be careful about complaining questions.

Secondly, watch out for contentious or combative questions. Questions regarding rules are almost always about changing the rule, not learning the basis for the rule. Teenagers are notorious for asking, "Why can't we…?" or "Why do we have to…?" Even some adults are consumed with asking contentious questions. Don't live your life as a question mark, always asking why or why not. **Live your life as a declaration of truth based on God's declared Word!** Remember, sometimes "good" questions can be foolish questions in disguise.

◇◇◇

DAY 340 — PREPARING THE NEXT GENERATION

II Timothy 3:17 That the man of God may be perfect, throughly furnished unto all good works.

We live in a unique time in human history. My grandpa farmed with a team of horses in Burdette Township, South Dakota. His grandparents lived in a sod house (actually a "dugout") on the banks of a creek in the middle of nowhere. In contrast, I am amazed when I think about what my kids have seen and experienced today: jet planes, computers, and other modern conveniences. There is a huge gap between my grandparents and my children!

Recently, I have been wondering what life will be like when my kids are my age. I can tell you that is a sobering question! If you have a toddler today, you will not leave a toddler to this world; you will leave an adult. The Bible says, *"And that from a child…."* in verse 15 and, *"That the man of God…."* in verse 17. What comes between *"a child"* and *"the man"* in this passage? The Word of God! *"All scripture is given by inspiration of God, and is profitable…."* (verse 16)

The key to leaving men and women who are equipped to face the future is to train boys and girls in the present. If you will do what you should right now, your children will be perfectly equipped *("throughly furnished")* for whatever lies ahead. God's truth makes the difference! His Word was sufficient for Timothy in the New Testament, and it is sufficient for you and your family today.

Every summer we have junior-aged boys and girls who ride horses here at the Ranch. How in the world does a ten-year-old junior boy who weighs seventy pounds control a horse who weighs more than nine hundred pounds? Simply put, you start training and leaving an impression on that horse when he is a foal (baby horse, for you non-cowboys). You don't wait until a horse is full-grown before putting a saddle on his back! You are just asking for trouble!

If you will prepare children, whether your own or ones to whom you are privileged to minister, you will produce able adults later. Prepare them **by example** (verse 10), **by explanation** (verse 15), and **by extension** (ministry opportunities outside your own home). By God's help, we *can* prepare the next generation to face whatever this world will look like in twenty years. We do that by teaching God's Word and by continuing in It.

"But continue thou in the things which thou hast learned and hast been assured of, knowing of whom thou hast learned them." (II Timothy 3:14)

THE BROAD VIEW

DAY 341

II Timothy 4:8 Henceforth there is laid up for me a crown of righteousness, which the Lord, the righteous judge, shall give me at that day: and not to me only, but unto all them also that love his appearing.

What is the most beautiful view you have ever seen in your life? Do you have something in mind? Without knowing your answer, I know something about your favorite view—it is probably a wide view. Whether it is a place in the Western U.S. or the panoramic skyline of New York City, **the best views are broad views!** There is no beauty when you are staring at the backpack of the guy in front of you; but after hiking into Grand Canyon, the broad view at the bottom is absolutely breathtaking!

What do see this morning? If all you can see is what is directly in front of you, you have a pretty sorry view! **There *is* a broad view for your life today.** Paul's immediate view was condemned (*"For I am now ready to offered"*), forsaken (*"For Demas hath forsaken me"*), and evil (*"Alexander the coppersmith did me much evil"*). However, **the broad view gave perspective to the immediate.** Paul had the broad view in mind when he said, *"Henceforth there is laid up for me a crown of righteousness, which the Lord … shall give me at that day…."*

When you have the broad view for your perspective, **you can leave your enemies, opposition, and troubles in God's hands.** Paul said of Alexander the coppersmith, *"The Lord reward him according to his works."* You can only have that perspective if you are focused on the broad view! Like Paul, you can confidently say today, *"Notwithstanding the Lord stood with me, and strengthened me…."* You can have vision, a good attitude, and the right perspective if you keep in mind the broad view.

How easily we get caught up in the "here and now" and all the things that consume our time, attention, and money. Don't be a slave to the view immediately in front of you; ask God to help you see the broad view. And the next time you see a picturesque view, remember that **the best view is the broad view!**

◇◇◇

DAY 342
LIVING WITH PURPOSE

Titus 2:1-4 But speak thou the things which become sound doctrine: That the aged men be … The aged women likewise, that they be … That they may teach….

I read a story some time ago about a man whose job it was to stand and watch the staircase in Parliament. For years, no one knew why he was doing this until someone discovered that his grandfather had done the very same job. His job was literally "grandfathered" in! Well, the story goes that years before this time, the grandfather watched the staircase when it had been freshly repainted. He was put there to make sure people didn't mess up the wet paint. The paint dried up, but the job never did!

Sometimes we can feel like that man standing guard at the staircase, knowing *what* we are doing but not *why* we are doing it. The book of Titus reminds us that **there is a big difference between knowing what and knowing why.** God has a reason and purpose for what He has asked us to do. Titus was to teach the *"aged"* men and women how they were to *"be in behavior."* Why was that important? They were to live godly in order to teach the younger men and women. **Living with purpose demands asking why before answering how.**

You cannot afford to live the Christian life without knowing why you are doing what you are doing. Knowing how and what are important, but you will not know these things with purpose without knowing why. You don't have to always know why in order to obey, but don't simply look for the list of do's and don'ts. **Success is not only accomplishing the "what"; success is reaching the "why."**

◇◇◇

WHAT KIND OF FRAME ARE YOU?

DAY 343

Titus 2:5b … that the word of God be not blasphemed.

My wife loves picture frames. In fact, we have boxes of picture frames with no pictures in them. I've learned over the years that you do not need a picture in order to buy a picture frame! For fun sometime, look in a fancy art gallery; and you'll, no doubt, see a tiny painting or sketch framed in a beautiful, ornate frame. The frame is a piece of artwork in its own right!

How you frame something is very important. What you see is often dictated by the way it is framed. This truth applies to artwork, your life, and the gospel. The way you frame the gospel in your life is important! You have the power to beautify or blaspheme the Word of God every day.

The word *blasphemed* found in Titus 2:5 means "to deride or to minimize." Paul was talking to Titus about both the older and younger men and women in the church. Their lives were to be clean so *"that the word of God be not blasphemed"* and *"that he that is of the contrary part may be ashamed, having no evil thing to say of you."* (verses 5, 8) **Does your life give ammunition to the naysayers of what is good, right, and biblical?** What is right and true is right and true, no matter how well you frame it; but people are less inclined to accept the truth if it is framed poorly.

My life is to frame the truth in such a way that the truth will be accepted and not rejected. I Timothy 6:1 reminds us that we ought to frame God's truth in such a way *"that the name of God and his doctrine be not blasphemed."* **There is no enemy of God or power in the world that can stop God's Word; but you can hinder It!** You can blaspheme God's Word by the way you frame it. God's pure truth given from a dirty vessel will only hinder and never help. The gospel is *"the power of God unto salvation,"* but the way we frame it is mightily important. When you see a beautiful frame around a picture or some artwork, remember that *you* are a frame for God's truth too!

◇◇◇

DAY 344
DO YOU LIKE ME?

Titus 3:2 To speak evil of no man, to be no brawlers, but gentle, shewing all meekness unto all men.

I've found that some people are easier to like than others. Have you found this to be true also? Showing kindness to the people we like is natural and requires no power; showing *"all meekness unto all men,"* as Titus 3:2 says, requires a power that you don't have!

I have had to remind myself that I'm not so likeable either! It is natural to like some and dislike others. On the other hand, it is *supernatural* to treat others the way God desires. **You can**

only show others what God has shown you. He has shown you a gentleness and kindness that is totally undeserved and unmerited. Only God can do that!

Showing meekness is not showing weakness. Weakness is weakness. A light touch when you are strong is meekness. It requires more power to restrain what you have than it does to exercise what you have. Showing *"all meekness unto all men"* requires a power which you can only have by living in the gentleness of God (verse 4).

Has God been gentle with you? Then you should show that to others. Has God forgiven you? Then you should forgive others. God wants you to "transmit" the power that He has. Take what God has shown you and show it to others. Showing all meekness to all men won't come naturally, but that is precisely the point.

MAINTENANCE REQUIRED

DAY 345

Titus 3:8 This is a faithful saying, and these things I will that thou affirm constantly, that they which have believed in God might be careful to maintain good works. These things are good and profitable unto men.

Do you like new things? Some people love the "new car smell," and no amount of air freshener can duplicate it! Some people love new pets—like my daughter's newly born kittens. Do you know what happens to cute little kittens? They become cats! And puppies are not any different! My nearly free chocolate lab dog has cost me an arm and a leg since we got her. **New things are exciting but demand maintenance,** whether you are talking about cars, cats, or relationships.

Often it is easy to feel revived and reenergized after a week of revival meetings, but the real work begins the moment the campaign concludes! The good things the Lord did during the meetings *will* require maintenance; change will not happen by default. A godly decision made at camp or in a revival service is a good start, but the work has only begun. We live in a wicked world and in sinful flesh, and both work against what's right. **Maintenance is required for any good decision you make.**

Titus 3:8 reminds us that maintenance is both required and constant. Paul said, *"I will that thou **affirm constantly**...."* That means every day! Start by getting your answers from the Bible, and then, with the Lord's help, get to work on maintaining good works!

I thank God for the good decisions that are made every summer here on the Ranch, but the real work begins when young people and adults leave the campground. Good decisions demand maintenance. Now if you'll excuse me, I've got some maintenance to attend to!

◇◇◇

DAY 346
OWING OR OWNING?

Philemon 1 Paul, a prisoner of Jesus Christ, and Timothy our brother, unto Philemon our dearly beloved, and fellowlabourer.

The relationship of Philemon, Onesimus, and Paul is one that reminds us of our relationship to God and to others. Paul called himself *"a prisoner of Jesus Christ"* and Philemon a *"fellowlabourer."* Paul asked Philemon to receive and treat Onesimus, Philemon's unprofitable runaway slave, like he would receive and treat Paul. You see, Paul saw Onesimus as a *"fellowlabourer"* because he did not think of himself on a higher level than this slave. Do you see yourself as a slave owner or a slave? Your answer to that question will determine whether you feel entitled or grateful.

It is easy to feel irritated, disrespected, overlooked, or wronged by others when you see yourself as "owning" and someone else as "owing." However, everything changes when you realize that you owe and God owns. Philemon was laboring under the impression that he owned and Onesimus owed. The fact was, Onesimus could have never owed as much to Philemon as Philemon owed to God!

Don't spend your day living under the impression that you own everything and others owe you. That is a recipe for frustration, conflict, and emptiness! Go into the day under the impression that you owe others and God owns you. **You don't own; You are owned.** An attitude like that will certainly make your day better, brighter, and beneficial to others!

IS CHANGE GOOD?

DAY 347

Hebrews 1:12b But thou art the same, and thy years shall not fail.

Who is the first U.S. President you can remember? I (barely) remember going to the polls with Mom and Dad when Ford and Carter were on the ballot. What is the first car you can remember your parents driving? My dad loves cars, and he can remember every car my grandparents had, every car he has had, and about half of the cars he has seen! Do you remember what the world was like twenty years ago? If you do, you would agree that our world has changed at an amazingly rapid rate!

Things are changing all the time! Some folks seem to like change, just for the sake of change; others have an aversion to any kind of change. **In a world of change, the only Person who *does not* change is God's Son.** The One Who was *"in the beginning"* is the only One who has not changed!

If you need an example of change in our world, just look at fashion. Fashion is so fickle! (Based on some of the fashions I've seen over the years, change is not a bad thing!) When styles change, at some point you fold the clothes up, put them in a box, and either give them away or wait fifteen years and pull them back out! For example, paisley ties like I had in Bible college years ago came back in style much later. People my age think my tie is "retro" while teenagers think I am cool and cutting edge!

At some point, any person you know, garment you wear, or car you drive will change. But unlike fashion, which always changes, the Lord Jesus Christ never does. He is better than the prophets that foretold Him and better than the angels that announced His birth. He cannot get any better than He is right now. **Good change in our lives is change that makes us more like the unchanging Son of God.**

DAY 348 — WORK ON RESTING

Hebrews 4:11 Let us labour therefore to enter into that rest, lest any man fall after the same example of unbelief.

Have you ever noticed how much *work* it takes … to relax? Planning a vacation takes an enormous amount of work! Vacation includes lining up someone to feed your dog, water your flowers, and watch your house. It's an oxymoron, but relaxing doesn't come without work. Likewise, resting in God does not come naturally; it takes work to rest in God.

Hebrews 4:11 puts together the words *labour* and *rest* in an intriguing way. *"That rest"* is God's rest, and only He can provide it. However, the Bible says, *"Let us labour…."* In other words, **resting in God requires work!** The Bible is full of examples of people who failed to rest in God. God had promised Canaan to Israel, but did they all conquer Canaan? No. Did they rest in God for the promise He had made? No. Some of the Israelites *died* in the wilderness because they failed to rest in God!

Are you the type of person who can fall asleep instantly? Some people can literally fall asleep at the drop of a hat; others have a whole routine they must do. Stores are full of items like hot drinks, little pills, and calming sound machines to help put you under! If it takes work to rest your physical body, doesn't it stand to reason that resting in God takes work too? How foolish to have promises from God but fail to rest in Him today!

Every promise of God is good because every promise of God is alive and powerful (verse 12). His promises for today are reliable; but just because He promised doesn't mean you will experience them. **Just because you *can* rest doesn't mean you *will* rest.** It comes down to an act of your will. You can either rely on the promises of God's Word, or you can run from them and rely on something or someone else. You can do much laboring today, but will you work to rest in God's promises? That is the key to true rest.

◇◇◇

ONCE

DAY
349

Hebrews 9:12 Neither by the blood of goats and calves, but by his own blood he entered in once into the holy place, having obtained eternal redemption for us.

In a revival service in South Florida once, a man indicated his need of salvation and spoke with me after the service about his need. He said, "Well, there are many interpretations of the Bible. You say one thing, and this denomination says another thing, and another religion says something different." He was confused by denominational doctrines and interpretations of Scripture.

That day, I took my friend to Hebrews 7:27, where the Bible says Christ *"needeth not daily … to offer up sacrifice … for this he did once, when he offered up himself."* If that were not clear enough, I took him to Hebrews 9:12 which says the same truth: *"Neither by the blood of goats and calves, but by his own blood he entered in once into the holy place, having obtained eternal redemption for us."* You cannot escape the fact that Christ made His sacrifice ***once***. It was a one-time event that has eternal consequences!

There is a difference between permanent and eternal. We have permanent ink and permanent press, but even the things we consider "permanent" are not eternal. Our focus is so narrow!

Even life itself is not guaranteed or permanent. Verse 27 reminds us that *"it is appointed unto men once to die, but after this the judgment."* **We all are guaranteed a one-time appointment with death (verse 27). But Christ's one-time act of dying for our sins carries eternal significance.**

Christ is the *one* way to God's *one* heaven, and His sacrifice was made *one* time. Even my friend in South Florida could not dispute what the Bible clearly says! If you have placed your trust in God's only Son as the only way to heaven, you can live confidently and securely today in God's salvation. A man may have great faith and be trusting in many religious things or people to get him to heaven, but if his faith is not in the right Person, he is sailing on a sinking ship! Our Savior paid our debt once and for all!

"For by one offering he hath perfected for ever them that are sanctified." (Hebrews 10:14)

DAY 350
THE REASON FOR CHURCH

Hebrews 10:24 And let us consider one another to provoke unto love and to good works.

If your immediate family is going to gather together during the day, where is the most likely place for the gathering? Inside your house, of course. If God's family in this part of the world is going to gather together, where is the most likely place for the gathering? Inside a church building. So why do we gather together at church?

"Well," you say, "church helps me grow and become a better Christian." Many motivations for going to church deal with us personally. Growing in the Lord, learning about the Bible, and obeying God's Word are all valid answers; but did you notice what the Bible gives as our motivation for going to church? *"Let us consider one another to provoke unto love and to good works."*

"Consider one another …" The Bible does not say to consider your duty, your growth, or your learning. The Bible says we are to consider others! **Going to church is not all about you!**

If every person at your church would consider other people, everyone would benefit, wouldn't they? That is the wonderful truth about giving. When you receive, one person benefits. When you give, two people benefit. When you give, the Bible says, *"It shall be given unto you."* The giver receives, the receiver receives, and the Lord is pleased. When you go to your church, do you leave the other people at your church better or worse?

God does say that we are to meet together, and the reason we are to meet is *"to consider one another to provoke unto love and to good works."* People should leave church refreshed because you were there. That's an amazing thought!

The primary question is not, what will you miss if you don't go to church? The primary question is, **what will *other people* miss if you don't go to church?** If they won't miss anything, that is a problem! You may never know how much an encouraging word from you may mean to someone at church. When you go to church prepared to give, you will receive. Church is not about you; it is about "one another"!

DAY 351
WHAT MOVES GOD

Hebrews 11:6 But without faith it is impossible to please him: for he that cometh to God must believe that he is, and that he is a rewarder of them that diligently seek him.

It's quiz time! Give one word that you associate with each of these names: George Washington, Abraham Lincoln, Henry Ford, Thomas Edison, and Bill Gates. You probably said President, honesty, cars, light bulbs, and Microsoft.

What words do you associate with Abraham, Noah, Moses, and Samson? You probably said father, ark, leader, and strength. Isn't that interesting? Almost every word you said described what they accomplished!

What do you think God notices about people? What would it take to impress the God Who spoke the world into existence? I'll give you a hint: it is not going to be anything that's accomplished. We see Microsoft, light bulbs, and a boat; God sees our faith.

God is not moved by what we do; He is moved by Whom we trust. Men see what I accomplish; God sees Whom I trust. If you accomplish all kinds of things but leave God out of the accomplishments, God is not impressed. Hebrews 11 is full of examples of people who accomplished great things, but they accomplished them all by trusting God.

As you focus on what you want to accomplish today, **begin by focusing on Whom you are trusting.** By trusting God and living by faith, you will accomplish what only He can do and please the only One who truly matters. People see what you do; God sees Whom you trust.

WHAT DO YOU VALUE?

DAY 352

Hebrews 12:16 Lest there be any fornicator, or profane person, as Esau, who for one morsel of meat sold his birthright.

Which would you value more, a porterhouse steak from a high-end steakhouse or a "daily special" steak? What if the porterhouse included an hour-long wait, but the buffet had no waiting? If you value good steak (and I do), then you probably want the porterhouse.

Now, let's say you skip lunch and are nearly starving when the scent from the buffet wafts to your nose? You are probably going straight to the buffet and scarfing down the "economy" steak, because at that time you value the immediate more than the long term.

The truth is, **the difference between pursuing the significant and pursuing the immediate is what you value most.** Do you value the valuable or the immediate? A person who primarily values the "now" is often called *profane* in the Bible. That simply means "common, base."

Esau is a glaring example of a man who valued the immediate over the valuable. He sold his birthright for a bowl of soup! So many times, we frame many of our daily decisions in the same way. Do you choose the option that buys you a little bit of time for now, or do you choose to invest yourself in what will be successful later?

Never sell what you can't afford to lose. As Evangelist Jim Cook says, the devil will give you what you want, but it will cost you what you have. Don't trade the valuable for the immediate! Make sure you value what God values.

WIL RICE IV

DAY 353

OUR CARES FOR HIS GRACE

I Peter 5:7 Casting all your care upon him; for he careth for you.

There seems to be a very fine line between caring about something and worrying over it. The more we care, the more we are tempted to worry. If someone says, "I'm not worried," what they really may be saying is, "I don't care." Well, there are things about which we should care. Maybe you care about someone who is far from God. Maybe you care about a financial need that is bigger than you.

That's why I love I Peter 5:7, *"Casting all your care upon him; for he careth for you."* That word *care* can mean a couple of things. It can mean either "I'm interested in something" or "I'm worried about something." There are things I need to take interest in, but be careful that your interest doesn't turn to worry. The more passion I put into things that are worthwhile and command my interest, the more likely I am to go off the side of the road into a ditch with worry.

The answer to this is not to care less. It's to worry less. That's what the word *care* means in I Peter 5:7. It's talking about anxious care, anxiety, or worry. **The Bible says that I don't need to care anxiously about my life because God cares tenderly about me.** He's interested in me, so I don't need to be worried about me.

Giving God my cares is sometimes an act of desperation. The Bible says, *"Casting all your care."* God's care of me is gentle, but sometimes the way I give my cares to Him is anything but gentle. It's like casting them at Him. It's more of a "God, help!" I throw my cares to Him, and that's OK because He cares for me.

I Peter 5:5 says *"…Be clothed with humility: for God resisteth the proud, and giveth grace to the humble."* The God that resists the proud is the same God that gives grace to the humble. Don't stop caring today, but live in humility by casting your cares to God. Know that He cares about you.

KNOWING AND GROWING

DAY 354

II Peter 3:2 That ye may be mindful of the words which were spoken before by the holy prophets, and of the commandment of us the apostles of the Lord and Saviour.

People know what they want to know. I can easily prove that to you this morning. Do you know who won the World Series last year? What is a carburetor? Do you know what the Rasmussen Poll is? If you don't know the answer to any of these, I probably have not tapped into your specific interest or passion.

Peter's intent in writing II Peter was to *"stir up your pure minds by way of remembrance: that ye may be mindful of the words which were spoken…."* So, what is your mind full of today?

Some people do not know what God has said because they are *"willingly ignorant"* and do not *want* to know. **Knowing is primarily a result of desire, not intellect.** When it comes to spiritual matters, you will know what you want to know!

You know what you want to know, and you grow where you want to grow. You have a choice to be ignorant or knowledgeable in the things of the Lord. Growing in your Christian life is a choice. Today, you are as close to God as you want to be. Be willing to listen to God's Word and to good people. God doesn't want you to stop knowing or growing!

"But grow in grace, and in the knowledge of our Lord and Saviour Jesus Christ."

DAY 355

IS YOUR FLASHLIGHT READY?

I John 1:5 This then is the message which we have heard of him, and declare unto you, that God is light, and in him is no darkness at all.

In the aftermath of Hurricane Sandy, 7.5 million people were left without power. Without power, the first noticeable thing missing was light. Unless you were impacted directly by Sandy, you probably didn't think twice about others without power because you had lights and electricity. The truth is, there are millions upon millions of people without God's Light. I'm afraid we don't think as much about that fact as we do about the aftermath of Sandy.

The Bible says that *"God is light, and in him is no darkness at all."* Our fellowship with God is characterized as *"walking in the light."* I John 1:7 says, *"But if we walk in the light, as he is in the light, we have fellowship one with another…."* A person without light cannot see, perceive, or go in the right direction. This person is literally "in the dark." **People without Jesus Christ have no power, no light, and no direction.**

The way you show the Light is to *"walk in the light."* Much like a flashlight gives light in a dark room, your life can show that *"God is light."* **The message those in darkness need to hear has been entrusted to those of us who have the Light.** Make sure your "batteries" are charged and you are ready to show the Light to a world in darkness.

WHEN HE SHALL APPEAR

DAY 356

I John 2:28 *"And now, little children, abide in him; that, when he shall appear, we may have confidence, and not be ashamed before him at his coming."*

I remember when I was ten years old and my dad drove our Chevy Citation X-11 from Jacksonville, FL, to Los Angeles, California. My mom and we kids flew a few days later to meet Dad in Los Angeles. I was shocked when I saw him waiting for us in the airport with four days of facial hair! I had never seen him unshaven before in my life, and I just couldn't get past it!

I was out of contact with my father for four days; so when we met at the airport in Los Angeles, I was caught off guard by his appearance. Often we think Christ's appearing will be like my experience at the airport. Christ's appearing will be an epic event, and there is no way to minimize that day. However, the event of His appearing should impact our lives in a real sense every day.

Christ ought to be appearing in your life every day, even before He appears in that day of His coming. That is your choice, not His. This reality is not merely a matter of "covering your bases" and "being a good Christian" so you won't get in trouble when He comes. Christ's daily appearance in your life has more to do with familiarity. In other words, if you get to know Him better every day, when you finally see Him face to face, you won't be ashamed.

Abide in Christ today. **If you are seeing Him in your life today, you won't be shocked if He appears tomorrow.** Those who know Christ but barely talk to Him, listen to Him, or give Him the time of day will be shocked and ashamed when He appears. It does not have to be that way for you, my Friend. When Christ appears at the appointed time, it will be more amazing than anything you have experienced before. But it shouldn't be different.

DAY 357

WHAT ARE YOU REVEALING?

I John 3:16 Hereby perceive we the love of God, because he laid down his life for us: and we ought to lay down our lives for the brethren.

What are you trying to hide? Some of us try to hide our age and wrinkles. Others try to hide their accent, no matter if they are a Yankees or a Southerner. And all of us try to hide our faults.

I John reminds us that **sin reveals the true nature of a sinner**, but **sin also conceals the true nature of a saint**. The new nature that God has made in me by His Spirit does not and cannot sin. There is a battle between what I was and what I am.

Love reveals God; hatred reveals the world. As God loves a world that hates Him, I should love people who do not appreciate me. If they are revealing their true nature apart from God, I ought to be revealing mine as God's child!

May God be revealed through my actions or words today! Don't just love in word; love *"in deed and in truth."* (I John 3:18) Your life should manifest Christ, especially in your love for others. Displaying Christ's love is impossible for you to do, but it *is* possible for the Spirit of God. Are you revealing Christ or concealing Him?

DAY 358

LOVE ONE ANOTHER

I John 4:7 Beloved, let us love one another: for love is of God; and every one that loveth is born of God, and knoweth God.

Have you ever met a stranger with whom you share a mutual friend? You probably felt like saying, "Any friend of his is a friend of mine!" That sounds warm and fuzzy,

404　　　　　　　　　　　　　　　　　　　　FIRST LIGHT

doesn't it? After all, your mutual friend has exceptional taste in friends, right?

However, what if after talking to this stranger, you find out that he is weird and your assumptions about "any friend of his is a friend of mine" were not true! Has that scenario ever happened to you? How could your friend like you *and also* like this clod?

The truth is, God loves many people whom I do not love. Now, if God loves them and I do not, He must know something that I have forgotten! **I am to love others because I am loved by God.** I should love because I am loved. People see God in this world when they see His love through my life (verse 12).

We love God when we love the people that He put here! You can't love God and hate your brother (verse 20). No one is more an enemy of God than you were at one time. **When you view yourself from God's perspective, you will see everyone else in the right perspective too.** Not one person should be loved by God but not loved by you!

"Beloved, if God so loved us, we ought also to love one another."

TO HAVE CHRIST IS TO HAVE LIFE

DAY 359

I John 5:11-12 And this is the record, that God hath given to us eternal life, and this life is in his Son. He that hath the Son hath life; and he that hath not the Son of God hath not life.

John Rice once told a story of a young man who was saved in a revival service one night. A few nights later, John Rice preached on the words of Christ to Nicodemus, *"Ye must be born again."* Again, the young man came forward, took John Rice by the hand, and said, "Brother Rice, the other night I trusted Christ to be my Savior, and now I want to be born again." Uncle John explained to him that when he trusted Christ, Christ saved him, and he got everything that God had for him. The young man replied, "Well, I wanted to be sure I had everything." John

Rice's conclusion was that when you have Jesus Christ you have everything in the way of salvation.

I John 5 says it this way, *"And this is the record, that God hath given to us eternal life, and this life is in his Son. He that hath the Son hath life; and he that hath not the Son of God hath not life."* Salvation is a gift. You don't earn it. You receive it or reject it. This gift is in a Person, God's Son, Jesus.

Whatever you may lack, when you have Jesus Christ you have life. Whatever you may have, if you do not have Jesus Christ, you do not have life. It's just that simple. John continues in verse 13, *"These things have I written unto you that believe on the name of the Son of God; that ye may know that ye have eternal life, and that ye may believe on the name of the Son of God."* So, we can know that we have life, and we can remember what gave us life to begin with—believing on the Lord Jesus Christ.

Assurance ultimately comes from depending on what Christ has done and not what I have done. Salvation comes down to what you do with Jesus Christ. That is the one crucial decision that determines death or life. It is the decision upon which we must hang our assurance. The way to have assurance today is to remember Who saves a person in the first place—nothing less or more than God's Son, the Lord Jesus Christ.

◇◇◇

DAY 360

HOSPITALITY: HELPING THE TRUTH

III John 8 We therefore ought to receive such, that we might be fellowhelpers to the truth.

It was dusk in a city many miles from my home. I had gone there without my wife, children, or even my dog. I was by myself. Quite frankly, I was a little lonely. Have you ever been there? Maybe you've been in the most familiar places of your life and still felt a little bit lonely. Increasingly, we live in a hostile world where we can be surrounded by other people, but still be lonely. The truth is, we need one another; we need hospitality. **God's plan for His children is that we be helpers to the truth**

by showing hospitality to those who walk in the truth. God's design for hospitality is for God's people to show comfort, encouragement, strength, help, and love to other believers.

III John is a book about hospitality. There are three kinds of people who are involved in hospitality: those who need it, those who give it, and those who refuse to give it. Who needed hospitality in this book? John did. John, writing to Gaius, said that he planned on coming to see Gaius face to face. He is talking about hospitality, and he is assuming that he will be granted hospitality by this brother, Gaius, when he shows up. We help the truth by giving hospitality to those who walk in the truth. There may be people around you today who have a place to stay and something to eat, but who still need hospitality. They need to be encouraged while they are seeking to walk in the truth.

Who gave hospitality? Again, this man named Gaius gave it. John says, *"Gaius, whom I love in the truth."* Truth is the purpose for this hospitality. John says, *"We therefore ought to receive such, that we might be fellowhelpers to the truth."* We don't show hospitality because someone went to our school, lives in our state, or because we just feel warm and fuzzy about them. We show hospitality because it's what Christ would do; and we help the truth when we show hospitality to those who walk in the truth.

Finally, there are those who neglect showing hospitality. This was Diotrephes. John says, He *"loveth to have the preeminence among them, and receiveth us not."* Someone who has put himself in first place has no room to show hospitality to others. Don't be like that! Put others first and help the truth by showing hospitality to those who walk in it. We are living in a hostile world, but God has provided everything we need to further the truth by encouraging those who seek to walk in it.

DAY 361

THE FIRST AND THE LAST

Revelation 1:17 And when I saw him, I fell at his feet as dead. And he laid his right hand upon me, saying unto me, Fear not; I am the first and the last.

I don't know about you, but when I come to the book of Revelation, I am slightly intimidated by and anxious of the events I read. What I don't understand can intimidate me, and what I do understand can make me anxious! The truth is, whatever you do or don't understand, the book of Revelation is *"the Revelation of Jesus Christ."* **Revelation isn't a revealing of things; it is the revealing of a Person.**

In Revelation, Jesus Christ is **revealed by His names.** For example, in verse 5 He is *"the faithful witness,"* the *"first begotten of the dead,"* and the *"prince of the kings of the earth."* No matter who is in power, Christ is sovereign. No one ever voted God in, and no one will ever vote Him out!

He is *"Alpha and Omega, the beginning and the end, which is, and which was, and which is to come, the Almighty."* He bookends all history and eternity.

Jesus Christ is also **revealed by His appearance.** John describes Christ in detail in verses 12-16. He describes the very God of heaven in all His majesty and power. What do all of these descriptions mean? While you may not comprehend all that you read here, the Bible is meant to be studied and understood, and this means the book of Revelation too! **You *can* understand the Bible, and you *can* know God.**

God predated every religion, every church, every preacher; and He will postdate every politician, leader, and king you can think of. He is everything! He is first. He is last. You don't have to understand every sign, every event, and every nuance of Bible prophecy to have confidence. If you know the Person the book of Revelation reveals, you can have confident assurance for today.

LOOKING AHEAD AND WITHIN

DAY 362

Revelation 2:23b …And all the churches shall know that I am he which searcheth the reins and hearts: and I will give unto every one of you according to your works.

You don't know the future, and neither do I. However, God does! Remarkably, God not only knows your future, but He also knows your heart. In the book of Revelation, before God gives us a look ahead at our future, He takes a look within our hearts. That is a good place for us to begin too.

Only God can know the future, and only God can discern my heart. Have you ever tried to discern someone's heart? It is fruitless and can be discouraging. We cannot know someone's heart because we do not even know our own hearts. The God that holds the future is the God Who can guide the heart!

God knows what is happening in our country, in our towns, and in our homes. Even more than that, He knows what is happening in our hearts. **Your choices today will be choices between what you want and what you should do.** Every time you choose what you want over what you should do, you have left your first love, like the church at Ephesus (verse 4).

Your life hinges more on the decisions you make than the decisions an official, a governor, or a President makes. God knows the future, and God knows you. Where is your heart this morning? The early church had God's power in a mighty way in spite of a hostile political climate. That is the power we need in our own hearts, our families, our churches, and our country. **God can see your heart, and He can guide your future.**

DAY 363

A GLIMPSE OF GOD

Revelation 4:2 And immediately I was in the spirit: and behold, a throne was set in heaven, and one sat on the throne.

Living in a fifth-wheel trailer presents, shall we say, "unique" experiences. One thing you quickly realize is that trailers are not sound-proof. You would think that no one else inside or outside the trailer can hear you, but that's not true! More than once, I've heard some nosey kids and adults on the outside discussing what it must be like on the inside. They might as well have been sitting on my couch talking to me, but they had no idea I could hear them loud and clear!

I've had those "I wonder what's inside that door" thoughts myself. I can remember passing homes in Beverly Hills that arrested my curiosity. Have you ever passed a palatial mansion like that and wondered what it looked like on the inside? I would love to just see what it looked like inside the front door!

Revelation 4 opens the door, as it were, to heaven and gives us a glimpse of the home of God Himself. There is no mansion, compound, or structure on earth that can compare to heaven! It is magnificent!

The greatest thing we will see in heaven is not a thing but a Person. John saw a throne *"and one sat on the throne."* This whole chapter is about the One who sits on the throne. What can we learn from this glimpse of God?

God is **holy**. The four beasts *"rest not day and night, saying, Holy, holy, holy, Lord God Almighty, which was, and is, and is to come."*

God is **omnipotent**. He is *"Lord God Almighty."*

God is **unchanging**. He is the Lord *"which was, and is, and is to come."*

God is **worthy**. The twenty-four elders worship the Lord and say, *"Thou art worthy, O Lord, to receive glory and honour and power...."*

God is **eternal**. He *"liveth for ever and ever."*

God is **creative**. He *"hast created all things, and for [His] pleasure they are and were created."*

Can you trust a God like that with your future? In a world of constant change, it is wonderful to know that God is capable, caring, and unchanging!

◇◇◇

ARE YOU WORTHY?

DAY 364

Revelation 5:12 Saying with a loud voice, Worthy is the Lamb that was slain to receive power, and riches, and wisdom, and strength, and honour, and glory, and blessing.

When you come to Revelation 5, one theme repeatedly echoes off the pages. You can't miss the word *worthy* in this chapter! Another closely-related word is *worship*. God is worthy, and He should receive worship. Someone is worthy when he has worth (value).

God is worthy, not because of what He does, but because of Who He is. That is in contrast to the way we normally view ourselves. Think about all the people we hold in high esteem and to whom we assign worth. We think we are worthy because of what we do. But we ought to worship the One Who is worthy because of Who He is.

So, what can we take away from all of this? Verse 10 says, *"And hast made us unto our God kings and priests: and we shall reign on the earth."* Does that mean we are big stuff? Well, no! **You will never find your worth in life until you assign worth to the One Who gives you life.** Christ is the One who is worthy! God made us; He redeemed us; and because of that, we are something prized in Him.

I am worth something because God is worthy of everything. Don't go through life trying to gain value by what you do. The more value you place on God, the more your life will have value. We have a Savior Who is worthy and should be worshipped. We are worth something when He is worth everything!

DAY 365 — ARE YOU A SUCCESSFUL MESSENGER?

Revelation 19:10 And I fell at his feet to worship him. And he said unto me, See thou do it not: I am thy fellowservant, and of thy brethren that have the testimony of Jesus: worship God: for the testimony of Jesus is the spirit of prophecy.

It is easy for a preacher to be so caught up with the message he gives and the notoriety it may bring, that he becomes enamored with himself. Sometimes, we think we give the message power instead of realizing that the Source of the message is also the One Who gives it power.

In Revelation 19, there was a messenger, an angel, who gave a fabulous message from God. John was so overcome by the joyous message given by this angel that he fell at the angel's feet to worship him. The angel said to John in verse 10, *"See thou do it not: I am thy fellowservant, and one of thy brethren that have the testimony of Jesus: worship God: for the testimony of Jesus is the spirit of prophecy."*

All of us should be giving the message of God to those around us. While giving the message, we must remember that **a successful messenger of God is a fellowservant who refuses worship while he reveals the Lord Jesus.**

John the Baptist had gained a great deal of notoriety, and he had disciples who were jealous for him because of the growing prominence of the Lord Jesus. In John 3:28, John quieted his disciples by saying, *"Ye yourselves bear me witness, that I said, I am not the Christ, but that I am sent before him."* John was saying, "My task is simply to be a friend of Jesus. I am to be a voice for the Word, a witness for the Light. I am simply a messenger." Whether the messenger is an angelic being or just someone who lives on earth, he needs to make sure that he remembers the power of the message is in the One Who gave the message, not in the messenger who delivers it.

God has given us a message worth giving. Remember that your success doesn't come when people applaud you; it comes when you are a fellowservant refusing worship while revealing the Lord Jesus.